A CLIMAT

We have most of the technology we crisis – and most people want to see more action.

But after three decades of climate COPs, our global emissions are worse than ever. And we are accelerating into a *Polycrisis* of climate, food security, biodiversity, pollution, inequality and more. What, exactly, has been holding us back?

What will it take for us to do better?

In the search for answers that have been so elusive to date, Mike Berners-Lee looks at the challenge from new angles. He *stands further back* to gain perspective; he *digs deeper* under the surface to see the root causes; he *joins up* every element of the challenge; and he *learns lessons* from our failures of the past. He uses all these insights to identify the single point of greatest leverage for those seeking the systemic change we so desperately need. *A Climate of Truth* spells out why, if humanity is to thrive in the decades ahead, the most critical step is to raise standards of *honesty* in our politics, our media and our businesses. Not only is this possible, but each of us can have radically more impact on the issues we care about by turning our attention to this simple principle.

Anyone asking, 'What can each of us do *right now* to help?' will find inspiration in this practical and important book.

***

Mike Berners-Lee is a leading thinker, researcher, best-selling author and consultant on the greatest challenges of the twenty-first century. About his first book – *How Bad Are Bananas? The Carbon Footprint of Everything* – Bill Bryson wrote, 'I can't remember the last time I read a book that was more fascinating, useful and enjoyable all at the same time.' His book *There Is No Planet B* was described by the *Financial Times* as 'a handbook for how humanity can thrive'. He founded and directs Small World Consulting, which helps organisations of every size and type to have a positive role in our world. Mike is a professor at Lancaster University, where his research includes emissions modelling, sustainable food systems and the impact of information and communication technologies.

'I was engrossed. It's now my go-to book. A hopeful and empowering book raising the challenges humans face and offering clear and straightforward actions for individuals, organisations and governments alike to take to reduce the worst effects of our planetary impact – all in bite-size chunks. A must-read for – well – everyone.'

Deborah Meaden, entrepreneur

'Reading this book is like diving into a freezing lake: it's bracing, lucid, and leaves you buzzing. Packing a life's worth of wisdom into every chapter, Mike Berners-Lee sets out the hard facts while holding the consequences with deep care. Dive in.'

Kate Raworth, author of *Doughnut Economics*

'Mike Berners-Lee's book is persuasive and informative. He is rightly dismayed at the lack of urgency in global efforts to limit climate change. This book will energise campaigners, and offer them, in compact and highly readable format, the information that citizens need.'

Martin Rees, Astronomer Royal

'The focus on truth-telling is crucial here. The author is right: if we can't cut to the heart of our dilemma, our chances of solving it are small.'

Bill McKibben, author of *The End of Nature*

'This is an astonishing, essential and radical book. I've never read anything that digs so deep and ties things up so completely. The brilliance of *A Climate of Truth* is that it cuts through the layers of mistruths to reveal the starker realities of the polycrisis we are facing. Everything that we do is entangled – and heading in a dangerous direction. Mike Berners-Lee provides effective solutions, but they are going to demand enormous changes in human behaviour, replacing our constant desire for growth with a genuine sustainable approach to living.'

Rosie Boycott, crossbench peer and climate campaigner

'Much has been written on the climate crisis and solutions to it, but few books are as comprehensive, readable and engaging as this one. Even fewer have something truly original to say.

'This book is different. Here, climate, social and political science meet behavioural economics, and more. The book covers values and value; expediency and honesty; law and lore; lying hard, lying down and taking a stand. A conversational style combined with serious erudition makes for ready reading, but this book is much more than just words. *A Climate of Truth* provides a real action plan that goes beyond the conventional list around the climate crisis with which many are already familiar. Far more than a moral sermon, it is a rallying call to action. Read this book.'

Hugh Montgomery OBE, University College London

'This is a book we all need to read – a humane, honest and intelligent approach to why we are where we are and how we can still make impactful change. Mike makes hope a generative action.'

Mary Portas OBE

'Mike's new book is a fabulous read, offering up tangible actions for every one of us and I thoroughly recommend it.'

Dale Vince, founder of Ecotricity

'Mike documents clearly why business-more-or-less-as-usual is not working and won't work. We need a Plan B. Mike provides the key components that we need for such a plan. Let's collaborate to co-create and implement such a plan.'

Pooran Desai OBE, founder OnePlanet.com and originator of *One Planet Living*

'The first half of the so-called "critical decade" was wasted on more of the same: denial, empty rhetoric and unrealistic technical fantasies about the future. This "age of lies" wasn't just driven by leaders incapable of addressing the climate and ecological crises, but also by deeply compromised experts and a compliant media. As we enter the second half of the decade, we've never been more in need of shedding the illusion that "others" will solve these existential problems for us. This is where *A Climate of Truth* comes in. In his unique style, Mike Berners-Lee dissects the challenges we face, explains why they remain unaddressed, and offers guidance for a new "age of truth". Crucially, Mike's contribution aims to empower us all with the tools to drive change ourselves. It's not about admiring the expertise of others but about equipping the reader to play their part in sparking a bottom-up transformation. Mike strikes a careful balance between opportunity and hope, without falling into the trap of ungrounded optimism. While I sense a greater urgency and need for social change than Mike suggests, I still wholeheartedly recommend *A Climate of Truth* to anyone looking to channel their concerns about the future into informed and meaningful action.'

Kevin Anderson, Tyndall Centre, Universities of Manchester and Uppsala

'Truth is an increasingly scarce commodity in a world of alternative facts and weaponised obfuscation. In the context of the climate and ecological emergency, this is more than frustrating – quite literally, a matter of life and death. As planetary breakdown spirals out of control, we are staring a grim, hothouse future full in the face, and only hearing and taking on board the truth will ensure we have any chance of spurring the wholesale action needed to stymie looming cataclysm. In this marvellous, inspiring and heart-felt narrative, Mike Berners-Lee reveals how embracing truth, honesty and plain speaking can not only transform every aspect of how we live our lives on planet Earth, but also ensure that our kids and their kids inherit a world worth living in.'

Bill McGuire, author of *Hothouse Earth: An Inhabitant's Guide*

'A tonic for our reality-avoidant political culture. Meticulous, measured and meaningful. You'll have heard of the audacity of hope: this is the audacity of truth.'

Jonathan Rowson, co-founder and CEO of Perspectiva, and author of *The Moves that Matter: A Chess Grandmaster on the Game of Life*

'This is Mike Berners-Lee's magnificent magnum opus. An excellent book distilling his work so far on the climate crisis and leading to the key issue of honesty, with chapters on how we can get truth into business, media and politics. This is all the more relevant given recent election results.'

John Bowers KC, Principal, Brasenose College, Oxford, and barrister

'Mike Berners-Lee shows us how the world's environmental problems – including the loss of biodiversity and climate breakdown – are all interlinked. He goes on to demonstrate that the obstacles to remedying these crises are not technological. Instead, he conclusively demonstrates that we need substantial social and political changes to cope with the growing threats. I was particularly struck by the emphasis he places on improving the standards of honesty in public life as a prerequisite for addressing the many challenges we face. As in all his writing, Mike is clear, forthright and compellingly persuasive.'

Chris Goodall, author of *Possible: Ways to Net Zero*

'Do you want to understand the "polycrisis"? Mike Berners-Lee takes you *inside* it. The crucial demand saturating this book is that – hard as it sometimes is – we have no alternative but to tell and face the truth. This book will provide you with crucial help in accepting this, and gives you vital information on what we all might do about it.'

Rupert Read, Co-director of the Climate Majority Project and author of *Why Climate Breakdown Matters*

'This is an extraordinary book that needs to become ordinary as soon as possible. Mike has developed a roadmap to take action in a polycrisis featuring nature and climate collapse. The care and respect he shows to both our psychological health and the science in developing his programme is outstanding. It doesn't matter if you've never done eco stuff before, this is what you do now. Thank you, Mike.'

Lucy Siegle, writer and producer

'This book contains both optimistic and uncomfortable messages. Climate change *can* be confronted. But for that to happen there's an urgent need to demand change in our dishonest public culture.'

Peter Oborne, author of *The Rise of Political Lying*

'A searing indictment of the disastrous impact that the calamitous decline in standards of truth and integrity in public life has had on our capacity to tackle the climate crisis.'

Richard Sanders, filmmaker

'This is a sobering yet compelling exploration of how politicians, media and industry have practised serial dishonesty about the crisis enfolding our planet. Unless we're truly honest about the irrefutable data we have, then the critical changes the world needs will never happen. Mike Berners-Lee makes an apocalyptic warning that profit and self-advancement is the most sinister pollutant of all.'
    Quentin Wilson, motoring journalist and transport campaigner

'Provocative, challenging, insightful and wise, *A Climate of Truth* is an essential handbook setting out the urgent changes we need to make. And chief among those isn't technological or even economic change: it's raising standards of honesty and integrity in public life. Berners-Lee's case is powerful and compelling: it is dishonesty that has been the biggest block to progress on climate for decades. And the most effective way to radically improve our chances of thriving in the years to come is by demanding honesty from business, from media and – most of all – from our politicians.'
    Caroline Lucas, former MP and former Leader, Green Party of England and Wales

'In a world in turmoil, besieged by war and threatened by climate change, Mike Berners-Lee demands that we reclaim this world for decency. *A Climate of Truth* is a call to arm ourselves against the liars and manipulators. It tears away the pretence and challenges us to respond. It is fundamentally disturbing because it is unrepentantly honest. Yet it is optimistic because Berners-Lee believes that we can and must stop the rot.'
    The Rt Hon Lord Deben, John Gummer

'Mike's rigorously detailed expertise is evident throughout ... an honest and insightful delve into how we must all strive to understand the reality of the climate crisis.'
    Tim Farron, MP

Mike Berners-Lee

A Climate of Truth

Why We Need It And How To Get It

CAMBRIDGE
UNIVERSITY PRESS

Shaftesbury Road, Cambridge CB2 8EA, United Kingdom

One Liberty Plaza, 20th Floor, New York, NY 10006, USA

477 Williamstown Road, Port Melbourne, VIC 3207, Australia

314–321, 3rd Floor, Plot 3, Splendor Forum, Jasola District Centre,
New Delhi – 110025, India

103 Penang Road, #05-06/07, Visioncrest Commercial, Singapore 238467

Cambridge University Press is part of Cambridge University Press & Assessment,
a department of the University of Cambridge.

We share the University's mission to contribute to society through the pursuit of
education, learning and research at the highest international levels of excellence.

www.cambridge.org
Information on this title: www.cambridge.org/9781009440066

DOI: 10.1017/9781009440073

© Mike Berners-Lee 2025

This publication is in copyright. Subject to statutory exception and to the provisions
of relevant collective licensing agreements, no reproduction of any part may take
place without the written permission of Cambridge University Press & Assessment.

When citing this work, please include a reference to the DOI 10.1017/9781009440073

First published 2025

Printed in Mexico by Litográfica Ingramex, S.A. de C.V.

A catalogue record for this publication is available from the British Library

*A Cataloging-in-Publication data record for this book is available from the
Library of Congress*

ISBN 978-1-009-44006-6 Paperback

Cambridge University Press & Assessment has no responsibility for the persistence
or accuracy of URLs for external or third-party internet websites referred to in this
publication and does not guarantee that any content on such websites is, or will
remain, accurate or appropriate.

For all those who think about the rest of the world when they vote, or are ready to start doing so.

CONTENTS

INTRODUCTION	*page* 1
How Did I Get Here?	4
How Is the Book Laid Out?	9
Before We Begin ...	11
1 THE CHALLENGE AHEAD	14
Breaking a Habit	14
Learning and Growing from Failure	15
An Evolution in Wisdom to Balance Our Technology	16
Just a Temporary Pause in Expansion?	17
Gaining Agency at the Global System Level	18
Truth	18
2 STANDING FURTHER BACK	21
Why Are Efficiency Gains Adding to the Climate Emergency?	21
Can Energy Efficiency Be Made to Help Us After All?	25
Why It Doesn't Work to Add Up the National Climate Pledges	26
How and Why Does the Economy Grow Differently from Trees, Mice, People and Elephants?	27
Do We Have to Grow Until We Pop?	32
The Layers of the Polycrisis	36
3 THE OUTER LAYER OF THE POLYCRISIS	37
Climate	41
Energy	52
Population	56
Food and Biodiversity	60
Pollution	66

Disease and Other Health Threats	69
The Technological and Physical Solutions	72

4 THE MIDDLE LAYER OF THE POLYCRISIS — 91
Why Is It So Hard to Look under the Skin? — 92
The Kübler-Ross Grief Transition Curve, and Why
    Is It Relevant to Us Now? — 94
The Broken Trinity of Politics, Media and Business — 98
Inequality — 102
Economics and Growth — 106
Law — 110
The Relationship Between Humans and
    Technology — 116
Education — 119
A Vision for the Middle Layer — 121

5 THE CORE OF THE POLYCRISIS — 125
The Inner Challenge — 127
Seven Thinking Skills for Tackling the Polycrisis — 128
Three Essential Values — 131

6 TRUTH – THE SINGLE MOST CRITICAL LEVER — 134
Why Is Deceit Easier Than Ever? — 136
What Do I Even Mean by Truth, Honesty
    and Deceit? — 137
Blending Fact and Fiction: The Perils of Bullshit — 140
How Can Honesty Win the Day When Lies and
    Bullshit Are So Easy? — 143
Why Do We Need Kindness Alongside Truth? — 146
Why Do We Need Strength Alongside Truth
    and Kindness? — 147
What Part Do Psychopaths Play and How Can
    We Contain Them? — 148

7 GETTING TRUTH INTO POLITICS — 151

- Is Dishonesty New in Politics? — 157
- Are the Results of the Post-Truth Experiment Coming In? — 159
- If We Let Things Slip Further, Where Does It Lead To? — 160
- Five Criteria for Assessing a Politician's Honesty — 161
- What Is Not Included in the Five Criteria? — 163
- What Can Politicians Expect in Return? — 164
- When Is It OK for a Politician to Change Their Mind? — 165
- How Should I Treat a Politician Who Has Abused Me with Dishonesty? — 166
- What Does It Take for a Politician to Regain Trust? — 167
- How Should I Vote If All the Candidates Are Unsuitable? — 168
- Why Is It Self-Defeating to Vote out of Self-Interest? — 169
- What Democratic and Parliamentary Reforms Would Help? — 170

8 GETTING TRUTH INTO THE MEDIA — 174

- How Can I Tell What Media to Trust? — 177
- What If I Enjoy Media That Fail These Tests? — 179
- How About Podcasts and Social Media? — 179
- A Tour of Selected Traditional UK Media — 181
- How Is the BBC Getting On? — 187
- What Can I Do to Improve the Media? — 191

9 GETTING TRUTH INTO BUSINESS — 193

- Truth in Marketing and Advertising — 198
- Truth in Consultancy — 200
- Truth in the Fossil Fuel Industry — 202
- Truth in the Other Extractive Industries — 203
- Truth in the Aviation Industry — 203
- Truth in the Food and Farming Industries — 204
- The Future of AI and the Tech Industry — 205

Truth in the Gambling Industry	206
What Does Greenwashing Look Like and How Do We Keep It in Check?	207
What If I Sniff Out Dishonesty in My Industry?	209
Are There Any Good Businesses Out There?	210

10 THE EVOLUTIONARY CHALLENGE AND WHERE EACH OF US FITS IN — 212

Have We Left It Too Late?	215
What Makes Me Pessimistic?	215
What Makes Me Optimistic?	216
Are Humans Too Fundamentally Selfish to Survive?	220
How Should We Deal with Crisis Anxiety?	221
When the Challenge Is So Global, What's the Best We Can Hope For?	223
How Can I Work Out What Part to Play?	225
What Can I Do? A Checklist of Actions	227
Finally …	232

APPENDIX 1 A TAXONOMY OF DECEIT — 234

Lies	234
Probable Lies	239
'White Lies'	240
Subtle Twists	241
Failure to Correct Errors	241
Staying Quiet	242
Loopholing	243
Misdirection of Attention	244
Biased Gathering of Evidence	245
Biased Selection of Evidence	246
Altering History	247
False Impression	247
Fake Judgement	249
Hidden Motive (or Influence)	249
Burying Bad News	250
Camouflage	250

APPENDIX 2 HONESTY AND TRUST CRITERIA — 252
Five Criteria for Assessing a Politician's Honesty – in More Depth — 252
Five Trust Tests for a Person or a Decision — 259

APPENDIX 3 WHAT DEMOCRATIC AND PARLIAMENTARY REFORMS WOULD HELP IN THE UK? — 262
Electoral Improvements — 263
MPs' Conduct and What They Should Be Able to Expect in Return — 264
Reform of Appointments — 267
Tools for Better Decision-Making — 268

APPENDIX 4 RESOURCES — 270
News — 270
Fact-Checking and Whistle-Blowing — 276
Campaigning Organisations — 281
Books — 283
Films/Documentaries — 293
Financial Information — 294

Acknowledgements — 296
Notes — 297
Index — 350

INTRODUCTION

This book is about the fact that humanity is accelerating into a deadly Polycrisis.

What do I mean by 'Polycrisis'? Alongside and inextricably linked to our climate crisis, we're facing an equally serious biodiversity crisis, a food security crisis for a rising population, a crisis of escalating, permanent pollution – and far more besides. The scientific evidence across all of these issues has been clear for decades, and, wonderfully, our problems turn out to be more or less solvable from a technical perspective. Yet, in spite of this, our response continues to be hopelessly inadequate. In fact, every single year, we degrade our environment by an even larger amount than we did the year before.

An increasing majority of us sense – at least on some level – that we are heading for a very dark place if we can't get to grips with all this. But despite our failure so far, and maybe even despite the truly awful re-election of Donald Trump, we may still change course, if we are fast, lucky and willing to put in the work. To do so, however, we need to change our approach – and in several ways.

We need to *stand back* much further from the problem to get the best possible perspective and insights into the big picture. And at the same time, we need to get much *deeper under the skin* of the problems to better understand their resilience and the reasons behind our persistent failure to date. We also need to *join up our thinking* much better, since every element of this Polycrisis of our own making is inescapably connected to everything else. We need to *learn* from our failures so far. And to do that, we need to *be truthful* about the difficulties we face.

Truth is essential to any progress we seek to make. It is the foundation on which all other progress rests. We have, somehow, to understand and overcome the lurches we have seen in the wrong direction. We will dive into this in more depth as the book progresses, but in short, we need to ensure the honesty of our data and the honesty of those in power, and to be honest with ourselves and each other about the situation we are in. We should neither overstate nor understate the challenges in front of us, nor the extent of our progress in dealing with them.

On this last point, I don't believe in being over-optimistic, but nor do I believe in being defeatist. Climate discourse is often polarised, from doom-mongering at one end to over-optimism at the other end, based on an (often technocentric) avoidance of the nature of the problem. Neither of these narratives will do. The former leaves us paralysed, and the latter leaves us complacent. I believe in being *realistic*, as this is the best way of reducing an issue from the nightmarish to the manageably factual: the best way of finding authentic *hope* and – in turn – motivating *meaningful* action.

> *Not everything that is faced can be changed, but nothing can be changed until it is faced.*
>
> James Baldwin, 1962

It may now be odds against us getting through the next few decades in good shape. But neither are we definitely doomed, because if we can get right to the poisoned heart of the problem we might, for the first time, be in a position to start getting somewhere. That is what this book is about. And, whoever you are, this book is also a guide to the targeted, high-leverage actions that *you* can take *right now* to be part of the health restoration that global society so urgently needs.

So, what actions are needed? And what's stopping us? Fundamentally, it's not down to a lack of scientific or technological ability. Yes, gaps in our knowledge still make things difficult in a few areas, but, as we'll see later, they aren't the bottleneck.

To answer these critical questions, we will have to peel back the outer layer of the problem and touch upon themes of politics, media, business, economics, investment, education, law and social justice, to see in very broad terms what we might need from them to be able to solve the technologically solvable challenges. We are going to see how acute and how provable the need for change is in these areas. Just as with the technologies for a sustainable world, the good news is that all the changes we could benefit from in the fabric of our society are physically possible – but again, that raises a deeper question:

What societal changes will it take for us to make the changes we need, in the ways that we run our society, to get on top of the Polycrisis?

We are going to end up looking at questions of honesty, truth, empathy, kindness, new ways of thinking, problem-solving and carefulness. We will explore why these things matter so much and how we can get them.

Since the challenge we face is urgent and practical, the end point, and ultimate purpose of this book, is to give the best and most grounded answer to the question:

What can each of us do right now to help?

While I've talked to as many wise brains as I can, I still don't have perfect answers. In fact, some of the questions we'll look at will only be answered with ideas, explorations and more questions.

But I still hope the suggestions will be empowering. You probably already know that cutting your carbon footprint is a good idea, but not enough to bring about the global systemic change that we so urgently need. You know we need better politicians, but your one vote isn't enough to deliver them. You may be feeling powerless in the face of a global problem in which you are just one small actor compared with mighty nations and companies. If that is the case, I hope this book, as a call to action, will leave you feeling that although you are just one person among the eight billion people involved in this struggle, your actions can have much more meaningful leverage than just flying less, reducing your meat intake and recycling your plastic yoghurt pots – even though it is still good to do those things!

How Did I Get Here?

Most people who know a little about my work think of me as a science-y, carbon number-y kind of guy. When I go on the radio that's what most people want to talk to me about most of the time, and all my books have been packed full of calculations and statistics.

Actually, I only got into that stuff by accident and somewhat to my own surprise. It came about as a result of asking: 'What needs doing?' and 'Where are the gaps?' ... and those same questions have prompted me to write this book.

My 20 years of working on climate started by addressing the most obvious challenges as they presented themselves. The journey since then has been one of digging down through the layers of the problem, to get to the core of the issue: to the heart of what is stopping us from taking action, and what it might take to finally begin sorting it out.

I started my work on climate change – as we called it then, rather than the more informative and realistic phrases 'climate emergency' or 'climate breakdown' or 'climate crisis' – by helping companies work out what they could do to drive down their own greenhouse gas emissions. I thought I would just do the strategic stuff and leave the carbon numbers to the people who specialised in that. But to my surprise, I quickly found that companies simply couldn't get hold of information about their climate impacts. So that is why I immersed myself in carbon numbers, embarking on a massive amount of work and research into supply-chain carbon accounting, in order to remove that obvious and basic barrier to climate action: you can only start to take action if you know what the problem is in the first place.

In 2010, I wrote *How Bad Are Bananas? The Carbon Footprint of Everything* (a fun but also deadly serious book about carbon footprints) to help bring about a more widespread and instinctive understanding of where the carbon was in our lives and in the world as a whole.[1] I also tried in that book to nod to issues under the surface, about how we live and think. It all felt useful up to a point, but it was inescapably clear that while a 'good enough' understanding of carbon numbers for individuals, businesses and governments might be necessary, it wasn't anywhere near sufficient in itself to cut our ever-rising emissions.

I'd begun thinking more about the system dynamics of climate, energy and gross domestic product (GDP) growth. A key moment came when Andy Jarvis at Lancaster University mentioned to me over coffee that he'd noticed the global carbon curve didn't just rise in a vaguely banana-shaped way but had actually been mathematically exponential for at least 160 years. It had a fixed growth rate of 1.8 per cent per year, give or take a bit of noise. My jaw dropped, just as his had done when he made the observation. Why wasn't everyone screaming about this? A mathematically exponential curve was incredibly unlikely to occur by accident. It strongly suggested something powerful was going on at the global system level that was impervious to anything humans had done so far to try to change its trajectory. It suggested that there was a fundamental process by which growth begets growth, energy begets more energy, and emissions beget even more emissions. It told us that somehow the system is correcting for any little piecemeal aberrations that we humans are throwing at it. It told us that humans had so far had zero – yes, zero – agency over the carbon curve. It told us something more was needed if we ever wanted to get anywhere at all to drive down emissions and change the trajectory of the curve, specifically:

> **We have to interrupt the dynamics of climate breakdown at the global system level.**

Andy wrote a paper for a top scientific journal, *Science*, and I got to work with journalist Duncan Clark in 2012 to write *The Burning Question: We Can't Burn Half the World's Oil, Coal, and Gas. So How Do We Quit?*, a book about the big picture of climate change. We wrote about the exponential curve and explored the system dynamics from a few different angles. One of the dazzlingly clear conclusions was that we had to leave fossil fuels in the ground. Duh! Honestly, how could such a ridiculously obvious conclusion have been so little talked about? But at the time it really hadn't been. While we had been writing, Bill McKibben published 'Global warming's terrifying new math' in *Rolling Stone* magazine and followed it up with his *Do the Math* lecture tour. Bill's article and his tour made the same critically important point go viral for the first time. In the book, we went on to point out that you wouldn't succeed in leaving all fossil fuels in the ground just by trying to persuade individuals, businesses or even a few countries to cut their carbon. In order to do so, and to repeat:

You have to interrupt the carbon curve at the global system level.

Over the next few years – and I like to think slightly aided by our efforts – the message about leaving the fuel in the ground became more widely understood[2] and is today grasped by all thoughtful climate commentators, even if it is somewhat skirted around or only vaguely mumbled by too many politicians, policy-makers, and even still by negotiators at the COP climate summits.[3]

But the carbon curve has carried on climbing, as it is still doing today, almost undented to the naked eye.[4] So to answer the question of what it might take to cut the carbon, I knew I'd have to stand even further back from the problem to see more deeply into it. What blend of political, economic and social change would it take for the world to start leaving that fuel in the ground? The question of how to get rapid enough change was becoming less technical and far more social. In fact, it was becoming increasingly clear that the technology – challenging though it is – was not the roadblock.

I also came to see that it wasn't useful to deal with the climate emergency in isolation. If we want to make any progress, we have to treat it as just one symptom of something much bigger that is going on for humanity. Climate breakdown is just one symptom of the difficulty we have in dealing with our own rising power as a species.

So, in 2018, I wrote *There Is No Planet B: A Handbook for the Make or Break Years*, which was an attempt to look at every aspect of the challenge in one short book. And in a way, the punchline of *There Is No Planet B* was a discussion of values and ways of thinking that humans now needed to adopt – like it or not – if we were going to thrive on our one and only planet, moving forward. I homed in on simple principles such as joined-up/big-picture thinking, future thinking, reflection, commitment to truth, and respect for all people and species.

Since then, five of those make-or-break years have passed, and despite a small blip for the COVID-19 pandemic, the carbon curve has carried on rising. The scientific community has issued ever-starker warnings that the worst symptoms of environmental breakdown are starting to emerge. Meanwhile, I've come to see that dishonesty, more than poor judgement, has lain behind most of the worst climate decisions. I've seen the re-election of Donald Trump make the dreadful dishonesties in UK politics and media look trivial. And I've witnessed the rising variety and sophistication of corporate greenwash. Continually, I have had to ask very carefully whether each project my own company (Small World Consulting) takes on is actually just part of the smokescreen. I've come up against increasingly cynical, sophisticated and highly effective misinformation from interest groups who fear sustainability will hurt their profits. I've contemplated the widening gap between the global cooperation that is required and current international relations, with two big wars breaking out in 2022 and 2023, and in 2024 the world's biggest military power relinquishing any meaningful sense of democracy.

I'm writing this book because – despite our pitiful response to date – if we make the right changes now, we might still pull through the dire situation that we are racing towards. With a following wind, there is even still a slim chance that our children could live better than we have done. But that won't happen through 'business as usual', 'media as usual' or 'politics as usual'. Nor will it happen by all of us waiting for *someone else* to do something. Nor will things change by using gimmicky sticking plasters, or the naive hope that technology plus the usual market forces will somehow avoid us having to tackle the root causes.

I don't know what the odds are of us avoiding billions of deaths and untold suffering. They are a lot less than 100 per cent because we have left it so late to be still accelerating into the crisis, but they are better than zero. That's all we need to know.

How Is the Book Laid Out?

As with most of my books, there will be lots of headings, many of them written as questions.

The first two chapters are going to be about the challenge ahead and the need to stand back to see the dynamics of the global economy and our current trajectory towards trauma from some relatively unusual angles.

The following three chapters will be a tour of the Polycrisis, layer by layer.

The outer layer will deal with the physical elements of the problem and its solutions.

The middle layer will look briefly at socio-economic elements of the Polycrisis, at the changes to our politics, media, business, education, legal system and so on that might be required for the technically solvable solutions to the outer layer to be put in place.

The core section will look in broad terms at the ways we need to think, the values we need to have and the qualities of decision-making that we require in the Anthropocene: the era in which humans are the most powerful influence on our now-fragile ecosystem. This will be followed by a guide to critical elements of decision-making and some tools to bring them about.

Chapter 6 looks at truth, which will have emerged as the single most critical lever for change.

The next three chapters explore very practically what we can all do right now to help bring it into the critical domains of politics, media and business: how we can evaluate all three, and what we can all do to raise the standard. Most importantly, this will be a push for a climate in which *truth* matters more than it ever has.

The conclusion sums up where we will have got to and what it means for us, standing back once more to see what we've found. There is a checklist of the high-leverage things any of us might think of doing to be part of the change.

There are also some chunky appendices which I think contain important material that didn't fit into the main flow of the book. They include some detailed thoughts on how the United Kingdom, in particular, can improve its political system through structural reforms, and a taxonomy of deceit, in order to flesh out all its many forms beyond simple lies.

Finally, the endnotes contain references to much of the foundational data that underpins everything. It is sometimes hard to demonstrate that an argument is robust within the space of a short book, but the references link to a wealth of stunning material and are there to be followed up any time you are wondering whether you can take my word for it, or simply if you want to know more. In the endnotes I have also sometimes fleshed out detail that I think is very interesting but doesn't quite fit into the main flow.

Everything I have written in this book about individuals, businesses and media outlets is backed by what I believe to be strong evidence. Sometimes I have been critical, and on these occasions I have often emphasised in the text, especially for the reassurance of my publisher's legal team, that I am stating opinion based on evidence that I find compelling, and I have referenced as much of this as has been practical. I also spell out in this book criteria by which I have been working out what sources to trust. Please accept my apologies for the use of phrases like 'in my opinion' even in cases in which the evidence I've presented speaks plainly for itself. I hope you understand the reason that I have resorted to such phrases, in a world in which the rich can use the legal system as a tool for bullying. I hope I have also treated everyone with the universal respect that I advocate, even in cases where I think that they have been grossly dishonest, disrespectful to the public and/or reckless with our world.

Before We Begin ...

Some caveats and words of reassurance: I'm not going to be encouraging anyone to get arrested, even though, as it happens, I'm supportive of that kind of action too if it's done very carefully and with absolute respect for everyone who is affected. It's not even about taking to the streets – although I've done that myself from time to time – and it's not about being a particularly strong person, or an unusually kind one, or even impeccably truthful, as I can't profess to be any of those things either. In terms of positioning, this is not a party-political book. In the United Kingdom, the United States and many other countries there has been appalling and routine dishonesty from a great many politicians in recent years. This cannot continue within any political party. I don't believe in 'false balance'; I'm going to pick my examples based on their illustrative value, rather than their political colour or out of any sense of duty to represent the whole political spectrum. The climate and ecological emergencies we face *must* transcend party politics, and in the end will require a huge evolution of how political systems function and how all parties conduct themselves. By the time you read this, the specific examples may be out of date, but the principles behind them are timeless – and in fact are becoming more essential each day.

The logic of the argument has taken me to writing quite a lot about politics, even though I don't really come from that world, except as an engaged citizen in an imperfect democracy. So you might ask, as I have asked myself, who am I to comment on such things? My answer comes in two parts. Firstly, the current state of affairs is so inadequate for today's situation that fresh eyes are essential, especially since the political sphere is so engrossing that for many it becomes myopic. All too often those with the most experience seem to have lost, or have never had, the ability to see politics other than through its own terms of reference. Secondly, although I have approached politics from first principles, I have augmented this with as much input from those with experience at the coal face as I have been able to gather.

Finally, given the importance of being truthful with ourselves and each other, I had better try to address my own biases. If I were somehow able to put aside the Polycrisis I would still, as it happens, have some personal and political inclinations, which would leave me more sympathetic to some politicians than others, even if I felt they were all equally honest. I'll mention some of them now, just for transparency and so that you can decide for yourself, as you read, whether they have any impact on, or bias, the arguments I'm going to make.

I think it is very important to reduce poverty and to enable opportunities for all. I think that requires a reduction in inequality, as well as the provision of high-quality essential services such as healthcare, education and a social safety net for everyone. I don't believe in the concept of 'trickledown' – the idea that making the rich richer is also a way of helping the poor. I do believe in taxing richer people, like me, who can afford it, as I believe that is, in the end, in everyone's interests and I do not believe in totally free markets. I think markets need to be adequately moderated by regulations and laws and, just as importantly, by public insistence on cultural values of fairness and equity. There is always room for differences of opinion, if they can be expressed truthfully and with an honest desire to learn from each other.

This is my attempt to be honest with you about where I stand, so there are no hidden agendas. As you read, please do feel free to take a moment to reflect on any biases that you may be bringing to your interpretation since, as we'll see, this is an essential habit for all humans to develop. Also, please do apply what you read here to your decisions about the politicians you vote for, the media you absorb, the businesses you buy from, what you do at work, how you live at home, and whether or not you join a campaign. **Your actions matter. They do make a difference.**

However bad things may or may not get along the way, it is only with higher standards of truth – higher than we've ever known – that we will finally stand to make progress on the biggest, and hitherto most frustrating, depressing and infuriating, issues of our time. My hope is that by the end of this book I'm going to give you a very clear sense of agency; and a sense of the simple but powerful things that all of us can do *right now* to make meaningful change. In that sense I aim to give you *realistic* hope – even if you have been losing yours.

1 THE CHALLENGE AHEAD

There are many ways of framing the challenge we face and the systemic reset that we need. Before going further, it is worth looking at this challenge from a few different angles.

Breaking a Habit

There are many, many people like me who are trying to help the world get to grips with impending environmental breakdown. But we have been stuck in a routine. We keep on working in our own different ways. We keep complaining at the failure of the world to listen properly and to act in a commensurate way. *The Limits to Growth* was first published in 1972 and *Silent Spring* even before that in 1962.[1] The basic, simple, indisputable messages behind these two seminal books have been screamed out by countless people and largely ignored for decade after decade. People like me, along with others who are bolder and more articulate, and those attempting more politically diplomatic language, make warnings. We point out that thresholds are being breached. We say the time to act is *now*. Then no action happens, and we complain again, saying that the time to act is *right NOW*. And we try to find a different way of describing the urgency that is even more emphatic than the words we used the last time around.

Although there are many different styles of approach, for all of us, it is time to admit that our efforts, our failures, our fruitless calls to arms and our renewed efforts, are a habit. Many of us have mental models of our whole lives playing out this way and imagine ourselves frustrated, angry and still complaining on our deathbeds as humanity spirals downwards. Maybe some of us half imagine a sour satisfaction in being able to say: 'I told you so.'

I don't want that life trajectory for myself. If it came to it, I'd rather have holidays in the sun while I can than waste my life fruitlessly complaining. But that isn't what I really want either. What I really want is to see the change we so badly and urgently need *actually starting to happen*. So that means spotting the dysfunctional cycles of failure and changing them.

If what you are doing is consistently not working, it is time to do something different.

Learning and Growing from Failure

Since nothing so far has delivered the change we need, many approaches are being tried out in the search for something that will cut through. That is good. Everyone has a different understanding of how the world works and what will be effective. People have written books, articles, academic papers, made films, become environmental consultants, personally modelled sustainable living, tried to set up exemplary sustainable businesses or become delegates at COP climate conferences. Many have taken to the streets or blocked roads, and some have even gone to prison. Some people say we have to work within today's economic mindset and framework because that's just the way the world has to work or because there isn't time to change it, while others say we can't get anywhere until we build a brand-new economics from first principles.

None of these things has so far delivered the change we need. Almost everyone trying to enable it thinks that some of the others working for the same ends are doing more harm than good. Nobody can really claim their approach is working, since *nothing* has worked.

Those pushing for a more sustainable world now need to stand back and learn from our failures so that they are not wasted experiences. We need them to become the building blocks by which we come to understand what it will take to actually make progress. We need to honour these heroic failures, and somehow stand on their shoulders.

To do this, we have to also peer under the surface of our failure to understand it. Firstly, of course, we do have to understand the most obvious and superficial reasons for failure. But then we need to dig deeper into the reasons behind the reasons. And then we need to dig still further to see the reasons behind the reasons behind the reasons. We have to get to the core of the problem. Only then will we be able to treat the causes of our failure, rather than the symptoms.

An Evolution in Wisdom to Balance Our Technology

Another way of looking at it is that we face an urgent *evolutionary* challenge to develop the governance skills that will allow us to survive alongside ever-rising power and technology within the confines of a finite physical space: our still mainly beautiful Planet A.

We often describe ourselves as a high-achieving species with plenty to be proud of. Tech-giant companies and tech billionaires describe their purpose as being to power humanity on to its next stage of achievement. An alternative perspective is that we are negative achievers, that technology without sufficient wisdom has made us the most destructive species on Earth. Our net contribution to the ecosystem has been worse than any other species. We have the most to be ashamed of and the greatest reason to humbly beg the ecosystem's forgiveness for our ineptitude.

1 The Challenge Ahead

Through our ingenuity and our technical brilliance, we have inadvertently brought a new era on ourselves. I'll use the word Anthropocene very simply to mean the era in which suddenly it is humans that are the biggest factor influencing the Earth's ecosystem. It represents a brand-new context in which to live, and it requires a new way of carrying ourselves. We haven't yet worked out what that new way is exactly, still less how to implement it, but we don't have any time to lose.

In the old context – the one in which we learned almost everything we know about how to do life – the world was a robust place. If we needed more of something, we could just expand. The level of our pollution was enough to affect small areas but not the whole world. We were unable to annihilate fish stocks, contaminate every corner of the ecosystem, change the atmosphere to mess up the climate, or poison or irradiate the entire world. Now we can do all those things and then some. We restrain ourselves from a few, but most we are doing at ever-faster rates. The change in context is that we can no longer get away with expanding our physical impacts.

After exploring what this Anthropocene is like, examining the consequences of not living well in it, and peeling back the layers of the problem, we'll be exploring what it will take for humans to be making the kinds of decisions that *can* allow us to thrive in this Anthropocene, how we can put that in place, and what each of us can do to help.

Just a Temporary Pause in Expansion?

Yet another way of framing the challenge is that we have to manage what might only need to be a temporary pause in the journey of human expansion. I titled my last book *There Is No Planet B* because for the next few generations, there won't be one – a 'Planet B', that is. I showed in that book with some very simple physics how utterly unfeasible it is, with the technologies that we currently know about, to bridge the gap to the nearest planet that could be any good to live on. We all hear about multi-billionaires who claim to be pushing at the frontiers of humankind's next great chapter of achievement – life in space. This may indeed one day be possible and even be an appropriate focus of attention. But right now it is so far away in any meaningful sense, and the crises we face are so pressing, that the space tourism hobbies of the over-rich have to be seen as an extreme dereliction of the responsibilities that come with wealth. The spare billions should be going into humanity's practical survival needs. We need to concentrate on making life work on our beautiful Planet A for the next century or two if Planet B is ever going to be an option in the distant future.

Gaining Agency at the Global System Level

We talk about the human trajectory of innovation and growth as if it is something that we control, but as we will see in the next chapter, if we stand back far enough, it is clear that so far it follows a predictable trajectory, over which humans have failed to exert much influence at all. We haven't needed to before, but we do now. For those of us who think in terms of free will, or live as if there is such a thing, the question is whether we will remain slaves to this trajectory, which ends in inevitable collapse, or whether we are capable of breaking free and exerting our agency. And if so, how?

Truth

As we look at each of the above ways of framing this challenge, a critical issue – perhaps *the* most critical – will come up again and again as we dig into the components of the Polycrisis. Whether your focus is on day-to-day practical politics or the crises facing our species, we can't get anywhere without higher standards of honesty.

How can we trust our politicians to act if they are dishonest about their vested interests in fossil fuel companies?

How can we weigh the potential and limitations of technology if companies fudge the figures to turn an ever-greater profit?

How can we cure ourselves of the Polycrisis if we won't even admit we're ill?

Can we even get by at all in the Anthropocene if our view is obscured by deceit?

We need radically more *honesty*, more reliably, in our politics, media and businesses, placing this issue at the very core of the Polycrisis. I want you to view everything you look at within this book through the lens of honesty, so that when we return to this topic in depth, you're aware of the stakes behind the 'mistruths' that plague our everyday.

One lesson we specifically need to learn from is the failure to uphold standards in public life. For many, the eruption of lying among many of our most senior politicians in the UK, US and many other countries was almost entertaining in the short term, but the consequences are already biting hard for some. Recently, we seem to have been experimenting in depriving ourselves of the qualities we need more than ever from our politicians, our media and our businesses. Although deceit has been a feature of these domains to some extent throughout history, the context in which we now find ourselves requires us to make an evolutionary step change. We can no longer afford to put up with it. That is why truth is one of the most critical leverage issues – or perhaps the most critical – for anyone who cares about the future. By pulling the truth lever hard, instead of loosening our grip as we have been doing, we can radically improve our chances of thriving in the Anthropocene. We will go on to look at how that can be done, but first we need to get a better perspective on the challenge by standing right back from the problem to look, in important but relatively unusual ways, at the physical dynamics of the crises we face.

2 STANDING FURTHER BACK

Humans have been getting more powerful, every year, for a very long time.

For millennia, the world was a robust and plentiful place for us, in which we could experiment and expand as far and as fast as we were able. It was a sturdy, bountiful playground, but now it is finite and fragile. In this chapter, we are going to stand right back to look at the *dynamics* of our growth, to see if they can tell us anything new about how – going forwards – we can live well on the beautiful, vulnerable home that we call Earth. It turns out that we will be able to make some fairly simple but widely overlooked observations, whose policy implications are so massive that no climate policy-maker can afford to ignore them.

Why Are Efficiency Gains Adding to the Climate Emergency?

Year-on-year, we get more energy-efficient at just about everything we do. It would be easy to assume that efficiency improvements should lead, by default, to a decrease in the total use of energy, and thereby a general decrease in the burdens on our environment. Indeed, under certain conditions that we *could* bring about, that *could* happen, but the default position, and how things operate right now, is in fact the exact opposite.

Why are our efficiency gains leading to us using *more* energy, not less as many people expect and assume? This phenomenon is sometimes known as the 'Jevons paradox', named after William Stanley Jevons, who noticed in the nineteenth century that as the UK became more efficient with its use of coal, it led to rising, not falling, coal demand and usage.[1] But the Jevons paradox doesn't just apply to coal. It describes a much more general and incredibly important principle:

> *When we find a more efficient way of producing or doing something, we usually increase the amount of it that we do by a bigger proportion than the efficiency gain itself. So the total usage of the resource and environmental burdens associated with it go up, instead of down.*

To unpick how this comes about, let's think of almost any process in our economy. Examples could include oil extraction, flying an aeroplane or making any household item. The process requires inputs (in the case of oil extraction, these include energy, materials to create oil rigs, pipelines, tankers, refineries, labour . . .) and it does things with those inputs to produce the useful outputs (oil) along with some environmental burdens (greenhouse gases, toxins and habitat degradation).

The *efficiency* of the process could be defined as the output per unit of input. So an efficiency improvement happens when someone finds a way of getting more out for any given amount that is put in. Let's see what happens when a process becomes, say, 15 per cent more efficient.

What we find over and over again is that the useful outputs become cheaper to make and to a higher standard. So they become better value for money, which leads to customers wanting more. It takes less energy to make and drive a car these days, but that has led to more and bigger cars being driven further.[2] This is also known as the *rebound effect* (Figure 1).

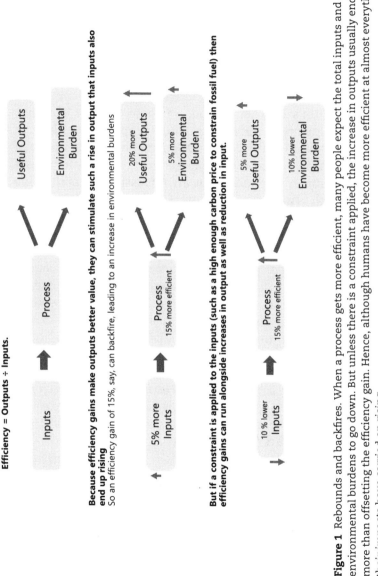

Figure 1 Rebounds and backfires. When a process gets more efficient, many people expect the total inputs and environmental burdens to go down. But unless there is a constraint applied, the increase in outputs usually ends up more than offsetting the efficiency gain. Hence, although humans have become more efficient at almost everything, their impacts have carried on rising.

The rebound effect is one of the most critically important and under-appreciated concepts for all climate strategists and politicians to get their heads around.

So far, I've described rebounds on a simple, single process, but in real life all these processes are part of a complex system that we call 'the global economy', and a myriad of direct and indirect rebound mechanisms take place involving interactions between different processes in the whole system. Let's look at just one of these, the car example: improvements in cars make some people more likely to live in the countryside, and with that comes a likelihood of having a larger home with a larger energy requirement. But these indirect rebound effects are impossible to capture one by one, so to understand their total impact you have to stand back for the macro view. If you don't do that, but instead try to understand rebounds by counting them up one at a time – as happens all too often – you vastly underestimate or even trivialise the enormously important rebound phenomenon.

The reason it matters so very much is that the overall rebound effect of energy efficiency in the global economy is currently, and always has been, *more* than 100%. Efficiency improvements are actually backfiring and leading to an *increase* in our access to, demand for, and use of, energy – including fossil fuel energy. So, coming back to our example, in the case of the global economy, a 15% efficiency improvement leads to something like a 20% increase in demand and therefore a 5% total rise in inputs, and their associated environmental burdens.

This is one way of explaining why global energy use and carbon emissions have been going up, not down, for the last couple of centuries, and continue to do so despite ever-improving efficiency of almost every process humans carry out: transport, heating, communication, data analysis, food production ... *everything*. (It is sometimes argued that because decoupling between energy and emissions has been achieved in some countries, this proves the decoupling concept at the global level. The problem with this argument is that it doesn't work to quantify rebound effects by looking at only one part of the system, such as individual countries.[3])

Here is an example of why you can't make useful decisions as a politician or climate policy-maker without a full grasp of rebounds. One of the outputs of COP28 was to push for a tripling in energy efficiency improvements. But because of rebounds, as things stand, this will lead to an *increase* in total energy demand and usage. And since we won't any time soon have enough renewable or nuclear energy to meet even today's energy use, the greater our energy demand, the more fossil fuel we will burn. The fossil fuel lobby at COP28 will have seen this very clearly. It has a far more sophisticated understanding of energy dynamics than the average politician.

Can Energy Efficiency Be Made to Help Us After All?

Yes. The dynamics don't have to work as I've described them. They can very simply be changed. All you have to do is *constrain the inputs* so that they can't go up. Then what you find is that efficiency improvements lead only to greater outputs and/or a reduction in environmental burdens. How you share out the benefit of efficiency gains between increased outputs and reduced inputs and environmental burdens is another matter, but the key thing is to understand the following:

> ***Only when the inputs are constrained do efficiency improvements stand to make both quality of life and the environment better rather than worse.***

Many in the fossil fuel industry don't want you to understand this because, in the case of energy efficiency in the global economy, it means constraining energy demand in order that efficiency gains can lead, for the first time, to reductions in carbon emissions. In practical terms, I think that means a high enough and increasingly universal carbon price, applied to the extraction of fossil fuel. More on this later.

Why It Doesn't Work to Add Up the National Climate Pledges

Failure to consider rebounds is also a fatal flaw in the international community's assessment of its carbon-cutting plans. The Paris Agreement resulted in a framework of national carbon-cutting pledges, the so-called Nationally Determined Contributions (NDCs). We hear that they are not enough to keep temperatures within acceptable levels, and that they are not all being implemented. But what is still missing from these assessments – and it is massive – is consideration of systemic effects. It simply doesn't work to think, as the United Nations does in its assessment of the NDCs, of each nation's carbon trajectory being independent of what goes on in other countries.[4] To do so is to assume, for example, that if a coal supplier finds it harder to sell its product to one country because of its carbon-cutting plans, that company will make no attempt to sell to another country instead. And if one country cuts its polluting manufacturing industries, the current modelling assumes this will never stimulate the relocation of those factories to other parts of the world that are less committed to climate action. In reality, if, as is the case in the UK, a country's climate pledges do not include emissions of imported goods and services, the tendency is to shift manufacturing to countries whose energy inefficiency and coal reliance may be higher, and in the worst case, actually lead to increasing global emissions.

I'm not saying we shouldn't have NDCs, but I am saying that much of the benefits of actions by individual countries will undoubtably be lost through leakage into other parts of the system, unless, as we will see later, something is done at the global system level to constrain emissions.[5]

How and Why Does the Economy Grow Differently from Trees, Mice, People and Elephants?

Now we are going to look at the growth of global society through a slightly different systemic lens, comparing and contrasting with other kinds of systems that we find in nature and in society. One way of looking at a tree is as a beautiful part of our environment. Another way is to see it as a complex system which carries out lots of processes. Its inputs are water, carbon dioxide, energy from the Sun and nutrients from the soil. It uses them to maintain its leaves, roots and branches and to transport all its nutrients to where they are needed. Its main outputs might include oxygen, fruit, and fallen leaves and branches. If there are any spare inputs after the basic work of staying alive and healthy has been carried out, the tree uses these to grow bigger.

It is a similar kind of story for most living organisms. All mammals eat food, drink water and breathe in oxygen. This gives them the energy to keep warm, move around, find more food, maintain their bodies, do all the things they want to in life and, if there is a surplus, to grow a bit bigger. A bit of physical growth is usually a good thing in children but usually not in adults.

As trees and many other natural systems grow, they find an economy of scale that allows them to keep getting bigger despite the fact that some tasks, such as transporting nutrients from the roots to the leaves, get disproportionately larger. Geoffrey West, former director of and distinguished professor at the Santa Fe Institute, has studied these growth dynamics to find some fascinating patterns.[6] He has identified incredible and uncanny similarities in the growth dynamics of different living organisms. For example, looking between all species of animals, you find that the rate at which they use energy goes up by almost exactly 75% every time weight goes up by 100% (Figure 2). The 25% difference between the weight gain

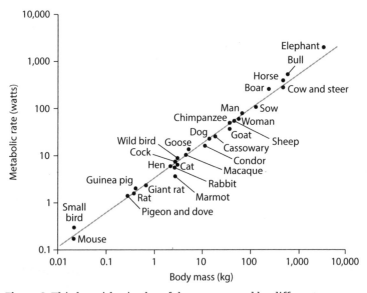

Figure 2 This logarithmic plot of the energy used by different mammals against their body mass is a remarkable straight line and shows that for every doubling in weight, the energy needed to live only goes up by 75 per cent.[7] (That is why the x axis increases by a factor of a million – from 0.01 to 10,000 – whereas the y axis only goes up by a factor of 100,000). Mammals are more efficient the bigger they are.
Figure credit: West, B.J. (2020). *Entropy*, 22, 1204.

and the energy use is the economy of scale that the larger species is able to find; in order to survive, bigger animals need to use fewer watts per kilogram of body weight. Elephants plod about, living longer but slower lives, with slow heart rates and with each kilogram of flesh needing less power than is the case for mice, who scurry around with heartbeats like drum rolls. Exactly the same ratio of size-to-respiration rate also applies to species of trees. (If you like maths, a ratio of ¾ between energy growth and energy use can be plausibly explained with a simple model in this endnote.[8])

Turning to the individual trees and animals, they also experience an economy of scale as they grow, but it becomes counter-balanced by moderating factors that inhibit growth. And because the economy of scale is a relatively modest 25 per cent, each tree, mouse, elephant or person reaches a point at which there is no longer any spare resource, and healthy growth has to stop (Figure 3). We call this maturity or adulthood and, once reached, the

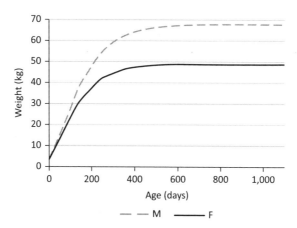

Figure 3 The growth rate of organic systems such as animals (in the graph here, sheep) reaches a stable plateau.[9]
Figure credit: Lupi, T.M. et al. (2015). *Animal*, 9, 1341–1348.

organism can continue to thrive for an extended period; perhaps centuries in the case of trees. In the case of humans, we can have many more happy decades, especially if we keep our intake in balance with our energy needs.

But West's work gets even more interesting and relevant to this book when he turns to exploring the growth dynamics of human social systems, cities, businesses and economies. His findings give us another powerful way of understanding the challenge facing humanity right now. In one sense they are alarming, but they also give us insight as to where our best hope may come from.

The key inputs to the global economy are energy, materials and food. We use these to enable all the activities and possessions that make up our lives, including the harnessing of more energy (from fossil fuels, renewables and nuclear sources), materials and food. When there is a surplus, this gets used for growth: growth in infrastructure and population, but also, unlike in trees and other mammals, growth in innovation and new ways of doing things.

Just like with plants and animals, socio-economic systems' growth and efficiency gains go hand in hand. But West found a key difference that has enormous implications for us. He found that in human social systems such as businesses, cities and even the global economy, when the system doubles in size, the rate of energy burn goes up not by a mere 75% but by a much larger and, again, amazingly constant 115%. So, whereas larger animals tend to have slower heartbeats and burn through less energy per kilogram of weight, in larger cities, the people live faster than in villages, and each individual uses more energy, not less.

In cities, compared with villages, people tend to walk and talk faster and, critically, interact with more people. As the global economy grows, it requires an enormous transport and communications effort to keep it functioning as a whole. So why doesn't it just run out of steam or buckle under the burden of such disproportionate energy needs?

The answer is that we have a whole different way of getting access to a greater energy supply. Trees, mice and elephants can't innovate, but they don't need to in order to grow to a certain size, plateau out and then continue to thrive. However, social systems both can and must innovate in order to find the extra energy that is required for the next stage of growth. In cities and companies, the rate at which ideas can be shared jumps up far faster than the size of the system. The same happens as our society continues to globalise. At the same time, we burn through far more energy per person than is required by a small, self-sufficient tribe.

Whereas trees and mammals naturally reach a point at which growth stops and the size remains constant, in human social systems, the innovation brings the efficiency improvements that power continuing growth. The result is a system-growth curve that – instead of levelling off at a point that we call maturity (in living things) – rises ever-more steeply. Growth begets growth begets growth, at an ever-faster rate. Energy use begets further rises in energy use. Innovation quickens. The more people there are, the more they can share ideas. The more spare resource they have, the more time there is to dream and experiment. It even becomes steeper than exponential. It becomes super-exponential. In other words, whereas in exponential growth there is a constant percentage growth every year, we are talking about that percentage annual growth rate *rising* as the years go by.[10] The more energy and materials we extract, the more we are able to do with them, including extracting ever-larger amounts, ever-more efficiently. The Stockholm Resilience Centre coined the phrases 'The Great Acceleration' and the 'Trajectory of the Anthropocene', to describe the eruption of steeply rising socio-economic trends. Figure 4 shows a whole range of socio-economic trends accelerating in this way. To repeat the phrase with which I started this book, we really are accelerating into a Polycrisis. But is it inevitable? Is there a way out, or are we doomed to failure?

Socio-economic trends

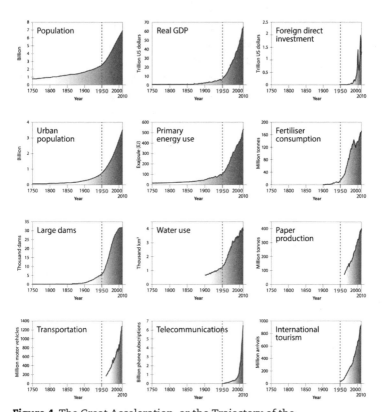

Figure 4 The Great Acceleration, or the Trajectory of the Anthropocene.[11]

Figure credit: Steffen, W. et al. (2015). *The Anthropocene Review, 2*, 81–98.

Do We Have to Grow Until We Pop?

Humanity has been experiencing the same dynamics of growth for millennia. We have been innovating and expanding. Small groups have been coming together into larger tribal units, then countries, unions and collaborations between countries, in a process that can be summarised as

2 Standing Further Back

globalisation. The energy supply has been growing. The bigger the groups, the more the interaction, and the greater the rate of that interaction. Energy begets energy, and innovation begets innovation. In our market economy, competition has been spurring everything on.

Geoffrey West describes a choice of two possible fates for socio-economic systems, of which global society is the largest and most important, that do not tend by default towards a healthy equilibrium state. The first is that they develop into ever-increasing vicious spirals and positive feedback so that the steepness of growth gets closer and closer to vertical: infinite growth (Figure 5). Because that is physically impossible to achieve (without expansion to

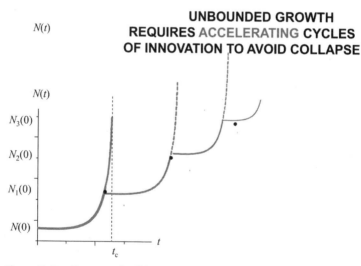

Figure 5 Geoffrey West's slide showing that socio-economic systems, of which the global economy is one, have a tendency to spiral into uncontrolled growth and inevitable crash, unless they undergo periodic resets. The need for these becomes more frequent until a meta-reset is required.

Figure credit: Graph used with permission from Geoffrey West.[12]

other planets, which as I've outlined isn't foreseeably feasible), the system is forced eventually into a dramatic crash, and often complete death.

The alternative, and what this book is arguing for, is that socio-economic systems are able to undergo a system *reset*; they manage to achieve a new mode of operation in which the rules and processes are different, and the growth rate is healthily tamed. Such a reset is one more way of framing the challenge we now face, as we are clearly not far from the asymptotic crash that West predicts.

In West's view, the growth rate eventually starts to pick up again after such a reset, and further resets are required, in fact with increasing frequency, until a meta-reset is necessary to avoid a meta-crash. But some wonderful news for us is that we can worry about this eventual meta-reset in years to come. Maybe by then Planet B really will be an option. In the meantime, we – humanity – need to perform a reset on ourselves – *now* – in order to survive. The alternative is that the planet *imposes* a reset on us against our will, leading to huge population loss, a crash of civilisation, an unthinkable level of collective suffering and untold destruction to the other lifeforms that share our Earth. Those really are the only two options: we either control the reset ourselves, to everyone's benefit, or we allow the reset to be imposed on us.

For now, in this practical book about the very present emergency of our inept and blundering arrival in the Anthropocene, the questions are 'What does that reset look like?', 'Who needs to do what?' and 'What can each of us do right now to help?' That will take us into questions about the ways we make decisions, the ways we think, to our relationship with technology, and for reasons we'll get into later, our relationship with truth. (Don't worry, we can still innovate – in fact we need to – but we need to have a lot more agency over the kind of innovation we embark upon.)

One of the great mysteries to unpick on our way to understanding humanity's failure so far to deal with the climate emergency is why, despite all the detailed assessments of climate impacts on every scale, there is so little attention paid to big-picture modelling of the system dynamics. This needs to change because even the relatively simple concepts explored here readily yield enormous policy implications. I don't claim to have the full answer, but I suspect it is partly, as we will explore later, because we haven't yet got used to thinking in some of the ways we are going to need to if we are going to thrive in the coming decades. It is partly because the implications are challenging and quickly make clear that minor modifications or sticking plasters on top of a global 'business-as-usual' approach won't be enough. It is also partly because, as we will see, some people are happy for us *not* to understand what it will take, for example, for us to leave the fossil fuel in the ground.

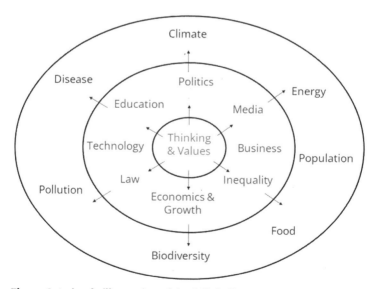

Figure 6 A simple illustration of the full challenge ahead: the Polycrisis.

The Layers of the Polycrisis

Over the next three chapters, we'll deconstruct the Polycrisis, layer by layer, starting with the outer crust and moving towards the centre (Figure 6). Then we'll look at the single biggest lever for achieving change, and explore very practically what can be done, and what each of us can do, to get things moving right now.

3 THE OUTER LAYER OF THE POLYCRISIS

Climate · Energy · Population · Food · Biodiversity · Pollution · Disease

We are going to look now at the physical aspects of the Polycrisis. These are the bits that are easiest to see and that will be the *direct* cause of our problems if we can't sort them out. For both these reasons, these physical challenges have attracted the vast majority of our attention. We are going to look at just how pressingly urgent the challenges are, how inescapably linked they all are, but also how wonderfully solvable everything in this outer layer of the problem is, at least from the perspective of science and technology. Unfortunately, that solvability should not give us cause for the optimism that some people take from it. Instead, it will force us to ask, 'Why *aren't* we solving these technologically solvable problems?' And that will take us to the next chapter which digs below this outer layer to look at what lies beneath. But before we get there, it is important to take stock of the physical realities, the looming threats and the opportunities, most of which we are not yet tackling.

Even if you have heard a lot about our climate and ecological emergencies, the next few pages contain perspectives that I think are not popularly grasped.

Warning: The next few pages are not going to be cheery. But we have to face this stuff fully if we are going to stand any chance of dealing with it. So, hold your nerve and read on. I don't think it is helpful either to overstate things for dramatic effect, or to understate the trouble we are steering into in order to make things sound more palatable. We need the best attainable view of reality (and the truth) that we can get. If we want to stand a chance of getting anywhere, we need to have the courage to stare this challenge in the eye. So be brave and be strong. Remember, this book does contain realistic hope and optimism – but only for a humanity that can face and honour the facts, as far as they can be ascertained. Remember also that you may counter-intuitively feel better after reading and acknowledging this stuff, because, like so many others, you already sense in your heart of hearts that things are badly wrong and heading in a worse direction. If so, your conscious acknowledgement will be a cathartic step. It might even come as a bit of a relief to know that a problem that was too intimidating to even look at properly has been downgraded to one that can be faced head-on.

Here are seven key components to the Polycrisis. I'm going to describe them as realistically as I can, starting from the top of the diagram (Figure 7).

The thickness of the connecting lines is intended to give some sense of the strength of connection between linked elements. Climate and energy are obviously totally inseparable because most (but by no means all) of our climate impacts come from our energy use. But food links to climate almost as strongly, since how we produce our food is also a huge chunk of our climate impact. As we will see, the way we produce our food is not only the single biggest driver of biodiversity loss, it also contributes strongly to our plastic pollution and the disease threats we face. Population relates to everything, since our total impact in every other area is the impact per person times the number of people. (For reasons that we will explore, population is not as strongly linked to climate as many people fear.)

3 The Outer Layer of the Polycrisis

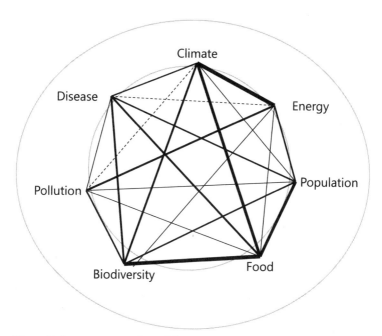

Figure 7 Seven components of the outer layer of the Polycrisis. Everything is interlinked, and the lines give some sense of the strength of connection between each pair of elements.

There are other models to describe this outer layer. In particular, most of the world has agreed to the United Nations' framework of the 17 Sustainable Development Goals (Figure 8). To me it is not a perfect model, partly because 17 is too big a number and partly because some of the elements are problematic: most notably, the conventional view of growth, which, as we will see later, is no longer appropriate in the context of the Anthropocene. Nevertheless, the Sustainable Development Goals are very useful because to some extent they give the world a common language to work from.

Figure 8 The UN Sustainable Development Goals.

> **The Sustainable Development Goals (SDGs)**
>
> In 2015, the UN created 17 world Sustainable Development Goals to ensure peace and prosperity for people and planet. The goals recognise the connection between environmental, economic and social aspects of sustainable development and specifically aim to address the issue of climate change and its impacts across multiple sectors of society.
>
> They are ambitious – initially intended as a framework for where we need to be by 2030, it is now widely accepted that many of the goals are unlikely to be met by then. Stumbling blocks include rising inequality, poor diet and therefore health, lack of investment from both governments and the private sector, and the lack of joined-up/bigger-picture thinking at both the national and global level.

Climate

Emissions Are Still Rising Even After 28 COPs

The science linking fossil fuels to climate change has been clear for many decades. Exxon's own models in the 1970s predicted the rate of warming with remarkable accuracy, even though their message to the public was to cast doubt on the science for many years.[1] Over the past three decades, the United Nations' Intergovernmental Panel on Climate Change (IPCC) has now held 28 huge international conferences – the climate COPs – to try to get on top of the problem. COP28 was attended by 97,000 people. Figure 9 shows us the extent of the progress we have been making on cutting the biggest driver of climate breakdown: carbon dioxide (CO_2) emissions from burning fossil fuel. The graphs give us the bottom line on how effective the COP meetings – and the resulting global and national targets and actions – have been. Looking at the curve you can see some dents from one-off events – such as the COVID-19 pandemic and economic crises – but you can't see with the naked eye any evidence that humans have even noticed climate change. The height of the line represents the *speed* at which we are making the climate worse through our burning of fossil fuel, so a rising line represents *acceleration*. In other words, we are still making the problem worse, and at an ever-faster rate. The most optimistic interpretation I can conjure up is that the long-term trend going back to 1860 was an exponential rise of 1.8 per cent per year, but this graph looks more like a straight line, which implies the *percentage* growth each year is actually falling.[2] Last year's growth (2023) was 'only' 1.1 per cent – arguably an improvement, but hardly cause for celebration. It's still growing.

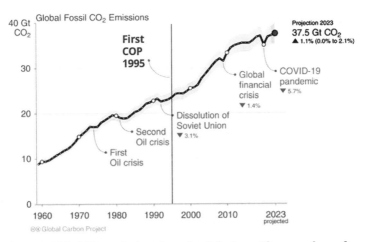

Figure 9 Global CO_2 emissions from fossil fuel use. The curve has a few dents representing economic events and most notably the COVID-19 pandemic, but it is very hard to see any evidence from this graph that humans have noticed they should be dealing with climate change.[3] Figure credit: Global Carbon Project.

I was recently told by a climate policy academic friend that she hates this graph because it suggests that we can't get anywhere and it makes people want to give up. I see it very differently. On the contrary, I put it in this book in order to maximise our chances of doing better in the future. This graph is a statement of fact. There's no point in wishing otherwise. Unless we come to terms with the reality that we haven't been getting anywhere, we don't give ourselves much chance of doing better.

Under-Rated Methane

While CO_2 is the most important greenhouse gas, about one-third of human-induced global warming to date has come from methane, and, in rough numbers, about two-thirds of the world's methane is human-induced. Of that, roughly a third is related to fossil fuels, about a third to livestock and about a third to everything else, especially waste (Figure 10).

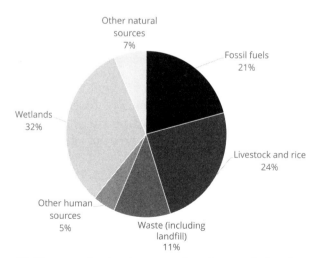

Figure 10 Human and natural sources of methane emissions in 2020.[4]

Methane and CO_2 behave differently from each other in our atmosphere. Carbon dioxide has a slow, steady effect on the climate. Each kilogram can be thought of as having a gentle warming effect but lasting in the atmosphere almost indefinitely. Methane, on the other hand, decays with a half-life of only around 12 years, but while it is up there, each kilogram of the stuff is warming the planet over 100 times more powerfully than its equivalent weight in CO_2. In climate change metrics, the widely held convention, for better or worse, is to measure methane's climate impact by comparing it to the impact that the same mass of CO_2 would have had over a 100-year period. When we do this, methane is said to have a global warming potential (GWP_{100}) of about 28 because each kilo of it has the same effect as 28 kilos of CO_2 over that same time period, having exerted nearly all its warming in the early years of the century since it was emitted. This method of doing the numbers is fine if the one thing you are interested in is understanding what the climate will be like in 100 years' time. However, if, as we need to, we also want to understand the impact over the next 20 years, then each kilo of methane emitted today suddenly becomes about 86 times as important as each kilo of CO_2 over that short time period. In other words, the usual way we measure methane leads us to under-rate, by more than a factor of three, the importance and critical benefit of cutting methane emissions as quickly as possible right now to give us faster results.[5]

Looking at the methane curve (Figure 11), it seemed to have been levelling off early this century, but of late has been rising more steeply than ever, with most of the recent rise coming from wetlands, as a result of climate change so far – the so-called wetlands methane feedback.[6]

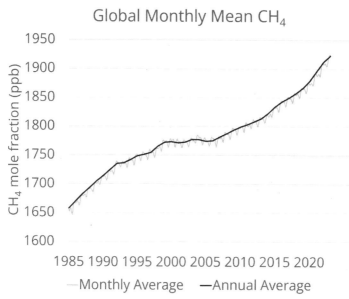

Figure 11 Methane levels are rising faster than ever, after looking to have stabilised in the early 2000s. Most of the rise is thought to come from warming wetlands. Melting permafrost and agriculture also play a part.[7]

Figure credit: NOAA.

A Multi-record Smashing Year

Now we've seen the emissions data, let's look at the effect it is having.

The 12-month period to January 2024 was the first to exceed 1.5 °C above pre-industrial levels.[8] This doesn't quite, in itself, amount to breaching 1.5 °C as defined by the IPCC, because the El Niño ocean current effect that comes around every two to seven years made 2023 a little warmer than normal. The long-term average temperature change is 'only' about 1.25 °C (2.2 F). Nevertheless, this 12-month period above 1.5 °C of warming has come earlier than expected and has been accompanied by a series of climate temperature records not just broken, but smashed by eye-watering margins – far more than can be accounted for by El Niño which, after all, is a frequent and repeating occurrence.[9]

The average global surface temperature record for September was broken in 2023, for example, by an astonishing half a degree Celsius.[10] Figure 12 shows just how unprecedented it is to have such a huge jump. In August 2023, the extent of Antarctic sea ice was not just a little bit less than ever before at that time of year, but a gigantic 11 per cent less. Figure 13 shows how dramatically 2023's sea ice area was smaller than at any other time in the past 50 years. Ocean temperatures change more slowly than land temperatures because our seas have such a high heat capacity, and they absorb much of the energy from global heating, dampening its impact. So when ocean temperatures broke records by 0.2 °C, that too was enormous cause for alarm. Once again, Figure 14 shows just how different the ocean temperature is from the previous record, which was set only the year before.

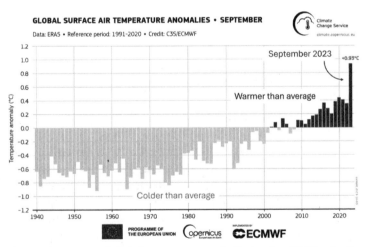

Figure 12 Average global air temperatures in September 2023 broke the record for that month by an astonishing half a degree Celsius.

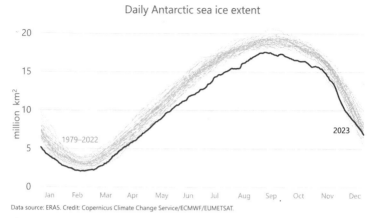

Figure 13 In August 2023, the area of Antarctic sea ice was an enormous 11 per cent lower than the previous record low for the time of year, which had only been set the year before.

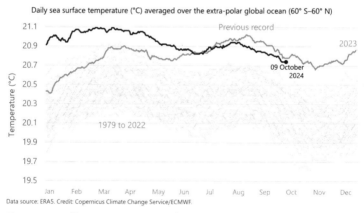

Figure 14 Daily sea temperatures have also been breaking records by very large margins.

We expect weather records of one kind or another to be broken from time to time owing to natural variations in climate and weather. If the world 100-m sprint record was broken by a couple of hundredths of a second, we'd be impressed and would perhaps want to be sure that the athlete wasn't on drugs. If someone broke it by a whole second – 10 per cent – we'd know that something was not right. Similarly, these enormous and unexpected jumps in the temperature and ice records have to be disconcerting.

Stop Press – Shortly before I submitted the draft manuscript for this book, the news came through that global average temperatures for February 2024 were 1.77 °C above pre-industrial levels.[11] The records just keep being broken.

Early Symptoms

These record-breaking pieces of data are one thing, but in real terms, why do they matter? The year 2023 was also a year in which we received early tasters of what the symptoms of climate breakdown are going to look like. While you can't attribute each extreme weather event to climate change, it is well established that the climate emergency is responsible for the rising frequency and severity of many such events.

Here are just a few of the examples from 2023. As you read them, bear in mind that *even with* very strong action on climate from now on, the Earth is going to continue to get even warmer, and even in the very best scenario, these kinds of symptoms are going to get substantially worse. That year saw wildfires in Canada alone adding 1.8 billion tonnes of CO_2 to the atmosphere. For comparison, the 1.1 per cent rise in global energy-related CO_2 emissions in 2023 relative to 2022 was only an increase of 490 million tonnes CO_2.[12] For a time, smoke from these fires made air quality in New York the worst of any city in the world. In the same year, wildfires on Maui in Hawaii killed 115 people and on the Greek Islands caused holiday-makers and residents to flee in their thousands. Heatwaves across Asia saw temperatures of 45.5 °C in Thailand, 44.1 °C in Vietnam and 52.2 °C in China. Beijing banned outdoor work after 27 consecutive days above 35 °C.[13] Europe and elsewhere also saw droughts, floods and heatwaves. Meanwhile, Afghanistan plummeted to minus 28 °C.[14] A record five million people caught the mosquito-borne Dengue fever, resulting in 5,000 deaths.[15] A record 42 ships made it through the Northwest Passage.[16] In 2023 we also saw early tasters of how the climate emergency is starting to affect food production and security. Hot weather cut Spanish olive production by 50% and global grape production by 7%. If you are thinking that we can live without wine and olives at a pinch, much more seriously, the price of rice surged in 2023 as India, which is responsible for 40% of global rice exports, saw yields drop,[17] and the global potato crop was also hit with climate-induced yield reductions.[18]

Self-Generating Climate Change

As troubling as all these early symptoms might be, with the sure knowledge of worse to come, what should be bothering us most is the poorly understood and irresponsibly under-reported risk of *tipping points*. It is easy to assume that the symptoms of climate change that we experience will continue to be proportional to the greenhouse gases that we put into the atmosphere; so if we double our contribution to the problems we might expect the symptoms to get twice as bad. If only that were the case. Unfortunately, there is potential for the symptoms of climate change themselves to trigger further warming – a **positive feedback loop** – and for elements of our ecosystem to reach a **tipping point**, beyond which the existing state is no longer viable and is destined to switch, regardless of how well humans start reducing their emissions.

> *It's the positive feedbacks and tipping points that should worry us the most.*

We've already seen the effect the wetlands methane feedback loop is having on the overall methane curve; how the temperature rise we have caused so far has already been enough for permafrost-trapped methane to explode out of the tundra, leaving craters up to 50 m wide. Methane is an extremely powerful greenhouse gas, which causes more global heating, to melt more permafrost and so on. To give another example of a positive feedback, rising temperatures have caused dramatic ice loss around the world on land and in the Arctic and Antarctic seas. Since ice reflects the Sun's heat, when it melts, the Earth absorbs more energy, triggering more warming (the albedo effect). To give yet another example, climate change is causing forests around the world to dry out, and in some cases die back or become more vulnerable to wildfires – causing more emissions and more climate change. To make matters worse, all these positive feedbacks occur at once and reinforce each other, leading to the uncomfortable idea – so far inadequately modelled – of **cascading tipping points**.[19]

At what point, then, might these feedbacks reach a level at which further warming becomes inevitable no matter how hard we humans work to cut our greenhouse gas emissions? The uncomfortable answer is that *we don't actually know for sure*. Until recently, this question hasn't had the scientific attention it deserves, and all the climate models have been based on a fairly linear response between greenhouse gas emissions and temperature change. As an illustration of how wildly we are throwing the dice, recent research suggests that the Atlantic Meridional Overturning Circulation (AMOC), which helps to bring warm waters to Europe, is likely to switch off at some unknown time in the next 70 years – perhaps as soon as 2025.[20] This would trigger rapid, huge and irreversible climatic disruption.[21] According to this research, Europe would be likely to see temperatures dropping by 3 °C per decade alongside serious reduction in rainfall and consequently dramatic reduction in food production.[22] This rate of change would almost certainly far outpace any of our attempts to adapt to the change (or any other species' attempt to adapt). Meanwhile, rainfall patterns in the Amazon could be completely overturned,[23] with equally devastating consequences to a part of the world that has central importance to the Earth's entire ecosystem.

In July 2023, UN Secretary-General António Guterres said: 'It is still possible to limit global temperature rise to 1.5°C [above pre-industrial levels], and avoid the very worst of climate change. But only with dramatic, immediate climate action.'[24]

Actually, Guterres was being over-optimistic. It is more accurate to say that we don't know whether or not we can avoid a catastrophic tipping point, and we don't even have a good sense of the odds. Scientists think that limiting global temperature rise to 1.5 °C will make it *less likely* that we will reach any of the major tipping points – but we don't know for sure. If we can stay below 1.5 °C of increase, then it still looks like we have a good chance of avoiding them – but only if we act *very* fast, *now*, to reduce greenhouse gas emissions.

Guterres also said, 'We have seen some progress.' Sadly, the evidence to the contrary is that we are making the climate worse by a larger amount each year. The fact that we are increasing the rate of emissions year-on-year means that we are still *accelerating* into the crisis, not just *heading* towards it. The first flicker of progress will be when the rate at which we are making things worse begins to slow. But there is no sign of that yet. In a survey of 380 climate scientist lead authors of IPCC reports since 2018, only 6% think the climate will stay below 1.5 °C, compared with 42% who think we will go beyond 3 °C.[25]

As I am writing, in early 2024, the world is experiencing an El Niño effect – a phenomenon of ocean currents that occurs every two to seven years, lasts for a year or two and has the effect of producing some short-term exaggerated warming. If and when we experience a catastrophic change, there is a good chance it will take place in an El Niño year. But if we get through the next couple of years, the immediate risk will recede a little until the next El Niño comes around, buying us some (incredibly precious) more time.

If, after reading all this, you can't believe in a pathway through these problems or see anything meaningful you can do to help, you may find yourself wanting to curl up in a ball. If so, stick with it, because while I can't promise to leave you feeling good about the situation we are in, and I won't give you a fake, cosy platitude, I do hope this book will take you to a place of realistic optimism, in which you can see a possible way through, and even more to the point, that you will be able to see a meaningful part that you might play. When that happens, I then hope that instead of feeling depressed, you'll feel an appropriate flow of adrenaline and be ready to take action.

Energy

Rising efficiency, rising demand, rising renewables and rising fossil fuel use.

What we need is rising efficiency, decreasing demand, rising renewables and steeply falling use of fossil fuels (to almost zero).

In March 2024, the International Energy Agency (IEA) reported on the latest energy trends: 'Total energy-related CO_2 emissions increased by 1.1% in 2023. Far from falling rapidly – as is required to meet the global climate goals set out in the Paris Agreement – CO_2 emissions reached a new record high of 37.4 Gt [billion tonnes] in 2023.' So far this is a sensible assessment of some bad news. But the contradictory gobbledegook that followed in the same report sums up some of the confused thinking that is so prevalent and unhelpful in so much energy analysis. The very same report said, 'Clean energy is at the heart of this slowdown in emissions' and 'The clean energy transition is continuing apace and reining in emissions.'[26] This is total nonsense. Firstly, emissions were not slowing down; they had accelerated by 1.1% to the highest rate ever. Secondly, there is no trace whatsoever yet of a clean energy *transition* because the renewables are not to any extent *replacing* fossil fuels, but merely supplementing them. The word transition means moving away from one thing to another, not having more of both.

Year-on-year we get more energy-efficient at just about everything we do. As I have already discussed, the counterintuitive reality is that this leads to us using more energy, not less. As we get more efficient with our energy use, unless we constrain *total consumption* in some way (such as through a carbon or energy price), we find that total consumption rises.

At the moment, the rapid rise in energy demand is being powered both by use of more fossil fuels and by use of more renewables. To give the renewables the faintest hope of replacing instead of supplementing the fossil fuel, we need to apply a constraint to fossil fuel extraction and use. The fossil fuel companies hate this idea. They want us to believe that we have ever-rising energy needs. We don't. While the poorest people and countries *do* have legitimate needs to access more energy, at the global level, we have falling energy needs, but rising energy *consumption desires*. The gap is *needless* energy consumption. Examples of this are everywhere but include: decreasing fuel efficiency of cars resulting from our appetite for SUVs; rising air travel; billionaires indulging in space hobbies; needless and pointless purchasing of material goods that don't add quality to our lives, whatever the adverts say; and the vast energy wasted though our failure to sort out the energy inefficiency of our homes.

Figure 15 illustrates, with three scenarios, the need to cut global energy use. Each chart shows total energy use to

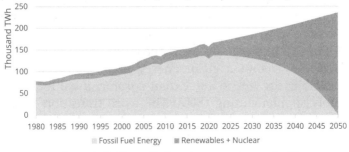

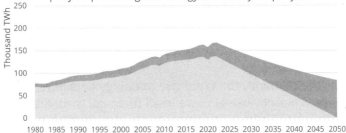

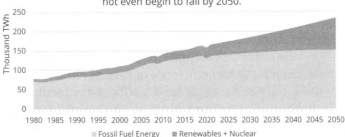

Figure 15 Three scenarios for global energy use, fossil fuel energy supply and non-fossil fuel energy production. Only by cutting total energy use (Scenario 2), contrary to the International Energy Agency (IEA)'s projections, do we stand the faintest hope of achieving net zero by 2050. Scenario 1 requires unrealistic renewable growth, and Scenario 3 – closest to where we are heading – is a disaster for humanity.

date, and projections to 2050, split into 'Fossil Fuel' and 'Renewables plus Nuclear'.

- In **Scenario 1**, world energy use follows the IEA's projection of about 2.2% growth per year to 2050 and shows that net zero by then is only possible if we increase Renewable plus Nuclear capacity by a massive 7.5% per year. That is an incredibly high growth rate, and it gets even harder to reach every year as the overall scale grows. Although net zero is reached, there is considerable overshoot of 1.5 °C because most of the emissions cuts are saved until the last minute.
- In **Scenario 2**, however, we bag the energy savings that are made possible through the continual efficiency improvements of all the world's innovation, instead of allowing these efficiencies simply to feed our energy growth dynamics. Thereby we cut global energy use by 2% per year. In this scenario the growth rate in non-fossil fuel energy only needs to be 4.1% per year to reach net zero by 2050. That is still a challenging renewables growth rate, but much more feasible. A further huge advantage is that compared with Scenario 1, the total amount of fossil fuel burned between now and 2050 (represented by the areas under the graph) is much less, because reductions start straight away. This means the temperature rise we experience on our way to net zero will be significantly less dangerous.
- **Scenario 3** is a complete disaster, but is where we are currently heading, at best. It shows what happens if energy use follows the IEA's projection and we still manage to grow non-fossil energy by 4.1%. Under this scenario, we don't even begin to cut fossil fuels before 2050. To repeat: it is a complete disaster for humanity.

Population

*** *Don't miss this relatively rare piece of good news in the outer layer of the Polycrisis* ***

Where Is Population Heading?

The ever-growing human population on this planet clearly increases the pressure on every aspect of the ecosystem. The more of us there are, the more lightly each one of us needs to tread. However, there is some under-reported good news coming up that population might plateau and then fall earlier than most people expect. Some economists and demographers have even expressed concerns that this rare piece of good news is something to be worried about, but we'll see why they are wrong to do so.

It took us roughly 300,000 years to go from a fledgling *Homo sapiens* species to reaching one billion in number. In a small fraction of that time, since 1800, world population levels have grown from one billion to eight billion – coinciding with, and likely enabled by, our burning of fossil fuels (Figure 16).

This has led some people to think, quite wrongly, that population alone is the root cause of all our problems; that it is the single killer issue. The idea of the human species being a foolish parasite that consumes its host has been around for a long time. Back in the eighteenth century, English economist Thomas Malthus predicted that human population would continue to grow until we brought about our own starvation. He argued that as we increased our supply of food, we would increase population levels at the same pace, leaving us no better off in the long run and indeed headed over the cliff of limited resources.[27]

Malthus's ideas still have a lot of traction today, even though the evidence is against them. For starters, he failed to predict the industrial revolution and, more importantly, how agricultural productivity would manage, so far at

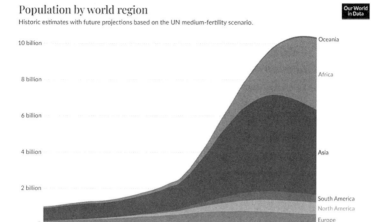

Figure 16 Eight-fold population growth since 1850 to date, with the relatively pessimistic UN central prediction of topping out at 10.4 billion in 2086. Excluding Africa, the rest of the world will peak much earlier, in 2050.[28]

Figure credit: Our World in Data.

least, to outstrip population growth (even though it has come at an unsustainable cost to our land fertility). We now have an abundance of food such that, if only we shared it around better, all eight billion of us could have a healthy diet, in spite of the gross inefficiencies of excessive animal products and avoidable wastes. Much more importantly – and this is very good news, as we will see later – Malthus didn't understand that people are *not* fundamentally driven to have as many kids as they are able to support in all circumstances. He was alive at a time when education was both expensive and non-compulsory. In England, in 1800, 40% of men and 60% of women were illiterate.[29] Many studies have shown that education, particularly of girls, has the strongest impact on moderating how many children a woman has.[30]

Indeed, when countries get above a certain level of prosperity (which usually goes alongside mainstream, compulsory education), their population growth almost invariably slows or stops.[31] South Korea, Japan, Sweden, the US, China and Spain are all seeing declining fertility rates,[32] with women having fewer children for a multitude of reasons. It is unclear how much of this is down to social factors and how much is, more worryingly, due to toxins slashing sperm counts.

Some very good news, just in terms of population predictions, is that the UN estimates (which most people use) don't take account of changes in birth rate with education, healthcare and life expectancy. Whereas UN demographers[33] expect the world population to peak at 10.4 billion in 2086 and to decline thereafter, some recent studies that have more fully modelled the effects of rising education and healthcare are much more optimistic (Figure 17). The most convincing study that I found was commissioned by the Club of Rome and carried out by Earth for All.[34] It estimated

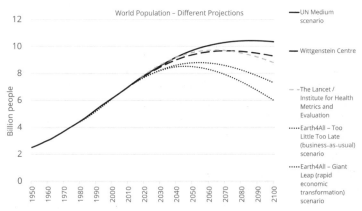

Figure 17 Widely varying population projections from the United Nations, The Institute for Health Metrics and Evaluation, the Wittgenstein Centre and Earth for All. The lower projections take fuller account of the effect of education, women's rights and poverty alleviation.[35]

that population is more likely to peak at 8.8 billion and as soon as 2050, followed by a rapid decline.

Whether through social improvements or as a result of a polycatastrophe, global population looks set to be falling soon.

A more dangerous, popular and misguided narrative is the idea that thriving societies *have* to have rising populations.[36] Some proponents of this idea say that almost no society in history has ever truly thrived while its population falls.[37] While this may be true historically, it is also true that past societies lived in a different context; one in which physical expansion was still possible. Others rightly point out that a falling birth rate means that in time there will be fewer young people to look after the old, but in doing so they fail to take account of several factors: the rising ability of older people to do many of the jobs that society needs; the rising efficiency with which many jobs can be done today compared with in previous generations; and the extent to which much of today's workforce currently does jobs that are not of net benefit to people and planet and could be scrapped to liberate workers for meaningful jobs in a society that needs more health and care workers. Traditional, GDP-based economic growth is facilitated by a young workforce, and on this basis, Elon Musk, Jordan Peterson and Jacob Rees-Mogg, all wealthy, privileged men who have 19 children between them, have urged others also to have more children,[38] on the basis that in a system of perpetual growth, more bodies are needed to support the 'weight' of those above them (the elderly). Those who argue that population growth is essential might usefully spend some time reflecting on where that takes us to in the end: 10 billion – or 20 billion – or 50 billion – or a trillion people on the Earth.

Ironically, the current economic growth model, combined with greed and too-low taxes for the over-rich, has led to the inequality that is causing many in the developed world to choose *not* to have children. The pathway we need to find is a well-managed decline in population, in which those who want to have kids feel they can afford to do so; no one has kids who doesn't want them; and the essential roles for a healthy and thriving society are filled with high-quality jobs. The reduction in the workforce can be met by having fewer jobs that lack intrinsic value to society – fewer 'Bullshit Jobs'[39] – and by efficiencies in the job market, even including appropriate use of artificial intelligence (AI).

Food and Biodiversity

The drive to feed an ever-greater number of people ever-more cheaply and with ever-more animal products has been hugely detrimental for biodiversity and land fertility, as well as for climate.

We've had a few decades of being able to bolster production through use of fertilisers and pesticides, but the result, exacerbated by the effects of global heating, is faltering soils, tottering yields and haemorrhaging biodiversity, including continuing deforestation. The inefficiency of feeding plants to animals and then eating the animals, compared with simply growing plants and eating them, is dramatically increasing the pressure. Techniques used to bolster animal product yields cause desperate environmental consequences, not to mention animal welfare nightmares. Meanwhile, hundreds of millions of us are under-nourished because the food industry doesn't enable nourishing, affordable food; hundreds of millions more are also obese because the food industry makes more profit from encouraging them into unhealthy diets. A select few make a *lot* of money in the food industry, but many farmers do not, and farmer suicides are soaring on both sides of the Atlantic.[40] It can't keep going this way, and it doesn't need to. We need to eat fewer animal products and grow food more wisely, with fewer pesticides and fertilisers. We also need to rapidly deploy the emerging science of agroecology.

> ### We Need to Talk About Chicken
>
> The production of every meat has its own issues, but chicken makes a case study that is indicative of the meat industry. The vast majority of chicken is grown through a system in which the birds have been bred to put on weight at the fastest possible rate – but that leaves them almost unable to stand. They are packed in tightly, because that is cheaper, and leads to fewer energy losses through chickens doing energy-wasting things – like walking around and having to keep warm. However, to prevent disease in this over-packed environment, and also to stimulate growth, they are fed antibiotics – raising the spectre of antibiotic resistance in humans. Antibiotics don't protect against viral infection. We know that bird flu jumping across to humans at scale would make COVID-19 look like a minor inconvenience. We also know these chicken factories cultivate exactly the conditions which exacerbate that risk.
>
> The streamlining of chicken production to minimise the cost per kilo of meat leads to a product that is nutritionally inferior to the product we were eating in the 1970s. The modern version typically has far more fat but only one-fifth of the omega 3.[41] And these chickens are fed entirely on grain, which requires vastly more land than would be required to provide the same nutrition to humans via a plant-based diet. Meanwhile, in the UK, for example, phosphorus pollution from the poo of the 20 million chickens in the Wye Valley is thought to

have almost wiped out the fish population in the River Wye.[42] It is possible to manage this excrement, but that costs money that currently not enough companies are prepared to pay. There are no wins here, and a lot of losses.

As long as you put it in a careful context, a bit of number crunching can sometimes go a long way to illuminating what is going on. Figure 18 shows estimates of the physical mass of different animals, made by Our World in Data. It gives a sense of how a rising human population and rising livestock numbers have pushed out the rest of the animal world. There is also now two-and-a-half times more weight in poultry than in wild birds.

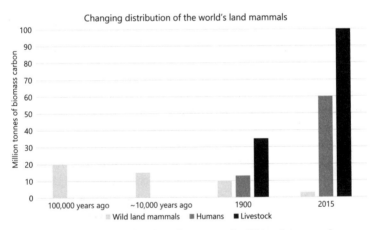

Figure 18 Estimates of the changing mass of wild land mammals, humans and livestock over time. By 1900, people and their farm animals were dominant, and in the years since then the picture has become even more extreme.[43]

Figure adapted from Our World in Data.

Sticking with numbers, here is a model of the current human food system, updated from some work I have previously published with colleagues at Lancaster University, and also written about in a previous book for exactly this reason.[44] It traces the world's calories, protein, iron, zinc and vitamin A through the food system on their journey from field to fork, and tells us a lot about what a sustainable food system might need to look like (Figure 19).

Humans require, on average, 2,358 calories per person per day to be healthy.[45] Figure 19 shows that we grow about two-and-a-half times what we need in human-digestible crops and that, on top of that, we grow more than twice the human calorific requirement in forms that are not fit for humans to eat, such as grass, pasture and some animal feed. Some of that animal crop is grown on land that could be used to grow crops for humans. Much of the rest is treated in a way that is terrible for biodiversity and climate because of a mix of over-grazing, monoculture crop production, use of fertilisers, pesticides and weed killers, and destruction of soil carbon through tillage and peat drainage. Some areas have been deforested specifically to grow more animal feed and for grazing. Nevertheless, there *is* a place for some land to be used, appropriately and sensitively, for agroecological animal farming.

With such a surplus supply of plant nutrition, you might ask how anyone could go hungry? What happens to all that food? To answer that, we have to follow the journey to human mouths (left to right on the diagram). Firstly, there are some losses in harvesting and storage, but that still leaves plenty of surplus. Then, critically, we feed a huge chunk of human-digestible crops to animals, and divert another huge chunk to non-food uses, which are mainly biofuels (which give us, in return, a minuscule proportion of our energy supply). Animals return to the food system only about *one-tenth* of the calories that they eat. They devour 4,793 calories per person per day (for every person on the planet) in grass, pasture and animal feed that, although deemed unfit for

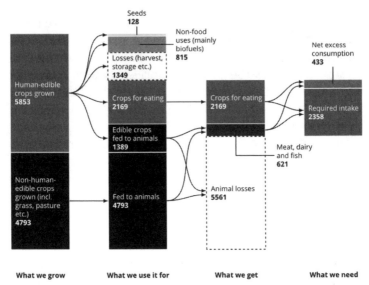

Figure 19 The global flow of calories from field to fork. Global calories (in kilocalories per person per day) on their journey from the field (left) to the fork (right), showing, most notably, the inefficiency of so much food going to farm animals, the current sufficiency of supply globally (if only we shared it properly), the various losses in the system, and the pressure that non-food uses (mainly biofuel) add to the food and land system. 'Animal losses' result, for example, from animals using energy to keep warm, move around and in some cases produce methane.

human consumption by the UN Food and Agriculture Organization (FAO), is largely digestible by humans. On top of this, they consume a further 1,389 calories in animal feed that the FAO classifies as fit for humans to eat. Animals return less than 590 calories per person to the human food supply, but this gets topped up to 621 calories when the wild-caught fish is added in. There are then some further small losses in transport, processing and distribution.

Eight billion end-consumers waste on average 245 calories per person per day, or just over one-tenth of the human requirement, although it is a lot more than that in some countries (including Europe and the US) and much less in some other countries. At the end of the day, humans eat on average 433 calories more than we require to be healthy, and this results in an obesity problem.

Of course, human nutrition is about much more than calories, so the next graph (Figure 20) tells a similar story for the world's protein.

Here you can also see the massive surplus production, though feeding crops to animals is not quite as wasteful of protein as it is of calories. The average human consumes nearly twice the protein they require to be healthy. The reason that not everybody has access to enough protein is that the market economy doesn't make it available and affordable to them.

For all the complexities of the global food system, these charts tell us a few simple and vitally important things:

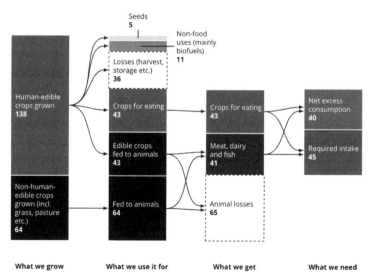

Figure 20 Global protein (in grams per person per day) on its journey from the field (left) to the fork (right). We produce plenty of protein, despite the animal losses, but we don't share it around properly.

- There is currently enough nutrition to go round if it were shared properly. The problem of sharing is that a market economy doesn't bother to reach out to poor people.
- We put huge and unnecessary pressure on the food system through the inefficiency of having so many farm animals; in particular, feeding them human-edible food, and using so much land for grazing and other animal feed. There is a place for farm animals, but many fewer and taking up less land for grazing and feed.
- It is obviously a good idea to reduce all the losses through the system as far as we can.
- There is substantial net excess consumption, especially of protein, even though not everyone has access to what they need.
- Biofuels also put pressure on the food system in return for a very small contribution to the human energy supply. (The term 'sustainable aviation fuel' is therefore a misnomer for bio aviation fuel.)

The whole food industry gives us another example of how the business world is not set up to maximise the health of people and planet. As well as unsustainable production practices, a great deal of food manufacturing and retail is geared up to maximise consumption regardless of health impacts. Obesity is both a cause and a result of overconsumption.

Pollution

Pollution of our air, land and sea takes many forms and causes more than nine million premature deaths per year.[46] Some forms are not as bad as they used to be and can be argued as success stories: relatively rare examples of humans reducing some of the negative impacts we are having on the planet. We are mending the ozone hole. Small-particle air pollution (soot) has come right down in many cities. Some of the worst pesticides are no longer used, and lead is coming out of our petroleum. But overall, the pollution story is not so rosy.

Three forms of pollution in particular are out of control. I've already mentioned that greenhouse gas emissions continue to rise even after 28 COPs. Secondly, agricultural pollution (especially fertiliser, pesticide and manure) is poisoning our soils, rivers, lakes and oceans to the point of threatening food production. But the steepest-rising major pollutant is plastic. This is especially worrying for its permanence, and for the difficulty in doing much about it. The plastic story epitomises humanity's current difficulty in turning down anything that is convenient or offers competitive advantage in a market economy. I'll discuss later the need for a radically new relationship with technology, and plastic illustrates the problem perfectly. We invented it. It turned out to be incredibly useful. So, in a free market economy we have found ourselves unable *not* to use it in ever-exploding quantities in almost every part of our lives, until suddenly we have an addiction that is very hard to tame, alongside rapidly mounting, permanent, environmental damage and health impacts.

The world now uses around 500 million tonnes of plastic per year, of which just 6% is from recycled materials and the rest is virgin, made almost entirely from fossil fuels.[47] About half of the 10 billion tonnes or so ever produced has been created in just the past 15 years,[48] and production has roughly quadrupled over 30 years. Almost a third goes into packaging; construction and vehicles together make up nearly another third of all uses; and the rest is spread across every kind of product you can think of.

About 20% of annual plastic production contributes to the stock of plastic that is stored up in the economy and in society, in everything from the plastic in buildings and cars, to boxes of children's toys gathering dust in attics. It is now thought to stand at over 400 kg for every human on the planet.[49]

The remaining 80% (around 400 million tonnes) of all plastic gets discarded as waste (Figure 21). Of this, around half (best estimate 49%) goes to landfill, where it stays forever. Nearly a fifth (19%) gets incinerated, capturing most of the pollutants if we are lucky and generating some energy while emitting CO_2 into the atmosphere. Roughly another tenth ends up in open-air dumpsites, which are ugly but contained, just about. We talk a lot about plastic recycling, but after many years of trying, still only a pitiful 9% goes down this pathway. About 7% is burned in the open air, with horrible toxins going up into the atmosphere along with carbon emissions, and none of

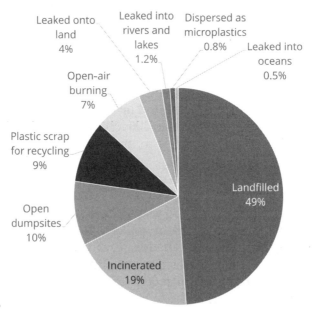

Figure 21 The fate of roughly 400 million tonnes of global plastic waste every year, of which only 9 per cent is dealt with acceptably.

the energy being captured. Around 4% ends up as scattered rubbish across the world's land area. Over 1% ends up in rivers and lakes, and about half a per cent in oceans. On top of all this, 0.8% is dispersed as microplastics, which get into every corner of the ecosystem: our soils, the plants, animals and fish that we eat, our drinking water and the air we breathe.[50] Once in our bodies, it gets everywhere: into penises, placentas, sperm, breast milk and the brains of unborn babies. We now know that the health impacts are serious. The phthalates (used to make flexible food packaging) and BPAs (used to make water bottles rigid) turn out to be disrupting our hormones (especially oestrogen and testosterone), dramatically cutting our fertility and triggering cancer.

So while some forms of pollution are on the mend, plastic is emphatically not. At around 5% per year, growth in plastic consumption is higher than fossil fuel growth has ever been and equates to roughly doubling the problem every 15 years. The plastics industry is lobbying hard for that growth to continue. Plastic is even more permanent than CO_2. As far as we can tell, future generations will forever be dealing with this rampant mess.

In terms of symptoms of the polycrisis, I have been slow to realise that plastic deserves an equal place alongside climate change and biodiversity loss. The plastic story also epitomises our current inability to live carefully on a fragile planet; our unreadiness to thrive in the Anthropocene.

Disease and Other Health Threats

Global life expectancy has never been higher. Our ability to prevent and cure diseases has been steadily improving, and machine learning (artificial intelligence, AI) stands to radically raise the pace of pharmaceutical developments. But we should not be complacent. As we saw earlier, some features of our food system are presenting serious risks. Because the population has both risen and increased its mobility, the speed of disease transmission also continues to rise. We will probably never know for sure whether COVID-19 came from a wet market, a laboratory or some other source. But we do know that some of our animal food supply chains are exposing us to risks, through the number of animals, the level of their contact with each other in crowded spaces and the extent of their contact with humans. If bird flu jumps to humans at scale (there are already increasing individual cases), with the potential to kill half of those infected, it will make COVID-19 look trivial, as I said earlier. Pandemics also threaten biodiversity. As I write this, bird flu, having already ravaged many wild bird populations, has recently arrived in Antarctica, where it is likely to rip through dense bird populations. In the US, it has been found in 20 per cent of retail milk samples, raising the possibility of widespread transmission between species.[51]

The World Health Organization is clear that antimicrobial resistance (including resistance not just to antibiotics but also to antivirals, antifungals and antiparasitics) puts the gains in modern healthcare at risk.[52] Some recent, encouraging scientific developments in new antibiotics are not enough to halt the trend of rising resistance. And we know that resistance is enabled in no small part by humans feeding two-thirds of the world's antibiotics to farm animals to help them grow faster and survive the unhealthy conditions and lifestyles that we usually enforce on them.

Both pharmaceutical companies and some food manufacturers currently benefit from the one billion people living with obesity, one aspect of which is an increased susceptibility to disease.[53] Ultra-processed foods may be as addictive as smoking, offering maximum instant gratification, but not making you feel full at the appropriate moment.[54] People who are obese metabolise more and therefore buy more food. Meanwhile in 2024, soaring demand for the anti-obesity drugs Wegovy and Ozempic saw Novo Nordisk hit a market value of $500 billion, making it Europe's most valuable company.[55] Big pharma also benefits from over-use of other drugs, including addictive painkillers and flippantly dispensed antibiotics – which, ironically, they sell more of the less effective they are. In many cases, the cheaper the food and the drugs, the greater the total sales value. Something is clearly not working optimally in the way these two industries serve humanity. We will see later, when we come to look at the role of business in the Anthropocene challenge, that many other industries similarly fail us. (The case of the Sackler family earning a fortune by addicting millions to the pain-killing opioid drug OxyContin gives us another example of how the pharmaceutical industry can make money from reducing rather than improving human health, even though here, clear-cut dishonesty – a major theme of this book – played a key role on top of perverse incentives for the health and pharma industries to have people in poor health.[56])

Climate change is already causing further disease threats, most notably by increasing the reach of carriers such as mosquitoes. Dengue fever outbreaks are increasing in some US states and have also spread to Spain. Other diseases likely to be on the rise from mosquitoes are Lyme disease, chikungunya and malaria. More generally, a study in the journal *Nature Climate Change* found that 58% of pathogenic diseases have already been aggravated by climate change, while just 16% have at times been diminished, and that overall, the rising impact of climate change on disease will outpace any adaptation attempts.[57] Heatwaves, storms, floods and harvest losses are other climate-breakdown-induced health threats.

Finally, we should not overlook the possibility of the deliberate spread of disease enabled by the same technological breakthroughs that are hailed for improving our capacity to cure illnesses.[58] It is not sufficient to know that AI can be used for health improvement; we also have to somehow know that it won't be used for nefarious means. It is wonderful that we have had so little biological warfare over the past century – a relatively rare example of human restraint in technology deployment – but it would be wrong to take this for granted in a world with so much conflict, and teetering on the edge of so much further escalation.

Overall, while disease threats are not quite at the centre of the Polycrisis, they form a significant part of the picture of threats on our doorstep. A quick look at the food and pharmaceutical industries has illustrated – and we'll return to this later – how the business world is not functioning to maximise wellbeing for people and planet, and the critical part that dishonesty plays in bringing this about.

The Technological and Physical Solutions

The main focus of this book is not technology, so all I'm going to do here is provide a flavour – just enough to make it clear that technology is not the bottleneck to a sustainable future. Plenty of others are writing in great detail about the wonderful technical solutions to our problems. There are some good books on this in Appendix 4, and I've even covered some of these solutions to some degree in my own previous books.[59] The fact that technological blocks are *not* what is stopping us will lead us, in a few pages' time, to look deeper into the problem. But first, here is a quick overview, starting with an incredibly brief summary, and then a little more depth, of those technical solutions that would be so heartening, if only they alone were enough to deal with the Polycrisis.

What Are the Top-10 Elements of the Technical Solution to the Physical Dimensions of the Polycrisis?

For all the complexity of the environmental challenge, a few simple principles will get us a very long way. To deal with the climate and nature emergencies we need three big things: a clean energy system, a sustainable food and land system, and a huge cut in pollution. So, the first four items on my list relate to clean energy, and the next four to sustainable food and land. They are both made massively easier by the final two items in my top 10, which are also essential in their own right: a circular economy (which is the most important measure for tackling pollution) and dealing with population growth.

Here is the list:

(1) **Reduce total energy demand.** The gap between total demand and renewable supply is met through fossil fuel, so this is the simplest and most important pillar of the energy transition. It doesn't get enough focus – for reasons that we'll come to.

(2) **Put a rapidly increasing price on carbon.** This is one tax that makes the vast majority of people better off. It scores several goals with one kick. It is the simplest and most effective mechanism for ensuring the fossil fuel stays in the ground. It incentivises energy efficiency. It provides revenue to simultaneously fund the green transition and cut poverty. It provides a dividend for low-carbon lifestyles.

(3) **Increase renewable energy supply.** This is vital, but remember that it only helps us if item 1 above is happening.

(4) **Develop storage and transportation of renewable energy.** Especially using hydrogen. This is crucial for dealing with the intermittent nature of renewable energy sources.

(5) **Reduce meat and dairy consumption and production**, not to zero but by a large amount globally,

and especially in developed countries. This is wonderful for climate, nature, food security and human health. It is technically simple and economically advantageous for all if it is implemented with respect and support for transitioning farmers.

(6) **Adopt agroecological production techniques** to reduce fertiliser and pesticide use, while enhancing yields and capturing carbon in soils. Some of this is technically complex and labour-intensive, but it is worth it and, with government support, stands to deliver rewarding livelihoods and thriving rural communities.

(7) **Develop lab foods.** Produce macro-nutrients (protein, carbohydrate and fat) using industrial processes that are renewably powered without the inefficiency of photosynthesis. In doing so, we can further relieve pressure for the highest possible yields at all costs on our farmland and also liberate significant land for nature and climate.

(8) **Produce nature instead of food** on land that can be rich in biodiversity but supports only low-yield food production.

(9) **Dematerialise our lives and circularise the economy:** extract less and more carefully. Buy less. Live with less clutter. Get stuff mended. Buy and sell second-hand. Share goods. Take the status out of shiny new things (from cars to gold rings). Improve recycling and quality of recycling, and reuse. Stop junk, especially plastic, getting into the ecosystem.

(10) **Enable population decline.** The key components are education, women's rights and poverty elimination. Then population looks after itself. The declining labour force won't be a problem if it is focused on just the jobs that benefit people and planet.

The top three on my list above come as a package. Renewables won't help much without a carbon-reducing energy demand. Both of these are enabled by a carbon price. Energy storage and transportation solutions complete the set for clean energy transition and get ever-more critical the further we get down the decarbonisation road. Note that direct air carbon capture and storage is important, but didn't make it into the top 10 and must not become a distraction from the key priorities.[60] This is because it is so limited in scope, and it won't help us at all unless items 1 to 4 are all working. Even thinking about direct air capture can so easily be, and already is, a distraction from putting them in place.

Items 5 to 8 are the keys to a sustainable food system. There is nothing not to like about the future we can have on this front. They will lead to happier, healthier people and a more beautiful, better-functioning planet. The biggest barriers are that there are some changes to get used to, some new skills to learn, and some powerful and cynical vested interests to contain. Government support is required to enable people to earn a living delivering a sustainable food system; there are some powerful corporations who will resist the change through all means available.

The final two items are last but definitely not least. Each deserves a list of its own, but I'm sticking to no more than 10 for now. As well as being important in their own right, circularising the economy and enabling population to peak and decline are huge enablers of everything else. They each help humanity to reduce its material, energy and food needs.

A Sketch of the Sustainable Energy Transition (Items 1 to 4)

Here is a whistlestop tour of items 1 to 4 in my top-10 list (Figure 22). We can argue about the details, but all I'm trying to do here is present a rough outline of what is entailed, to demonstrate the broader point that it *is*

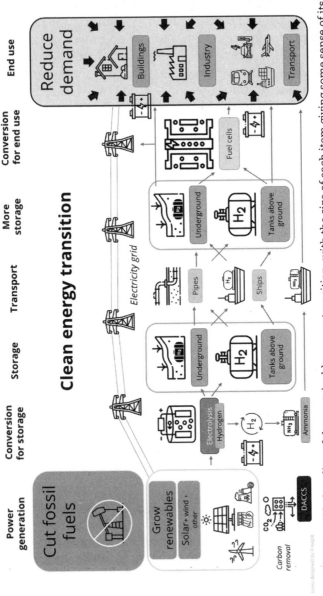

Figure 22 A simplified outline of the sustainable energy transition, with the size of each item giving some sense of its centrality – although we need everything here. Note the biggest boxes are for 'Reduce demand' and 'Cut fossil fuels'. Also note the relatively small size of DACCS (direct air carbon capture and storage).

possible, and to highlight some of the key elements. The size of the shapes is roughly related to how much we need to focus on them.

Demand reduction is the most critical and under-discussed component of the drive to leave fossil fuels in the ground. The fossil fuel companies hate the idea of us using less energy, and they work hard to protect us from understanding the clear-cut need to do so.

Reducing our energy demand will come from a blend of infrastructure investment, innovation, changing consumer habits, policy incentives and cultural shifts – each one of these elements supporting the others, and not waiting until the others are in place before getting started.

Buildings need, of course, to be energy-efficient, especially using insulation, draught reduction, mechanical air heat recovery, and heat pumps for both heat and, increasingly, for cooling. Where possible they need to generate and store at least some of their power on site, thus relieving the pressure on peak demand. (The UK's leaky building stock will be a growing lead balloon on the economy, passing down through the electoral cycles, until some government or other invests in it. Gas currently supplies around three-quarters of UK domestic energy, so shifting to electricity instead will require a huge upgrading of the electricity grid, even if there is a big improvement in energy efficiency.)

In particular, the steel industry is ready to transition from using coking coal to hydrogen, and alternatives to traditional cement are also on their way.[61] Every industry, obviously, needs to become as efficient as it can and circularise, producing less but better. Repair, maintenance and recycling industries all need to grow and improve their quality.

There is no getting around the extremely high energy use and emissions from **aviation**. It accounts for around 2.5% of global CO_2 emissions and rising, but aviation has a much higher contribution to global heating because of so-called high-altitude 'radiative forcing' effects, around which there is some uncertainty, but a central estimate is that they increase the climate impact, making aviation about 3.5 to 4% of all human-induced warming.[62] The technologies for sorting this out are a long way off. Planes have become more than twice as efficient per passenger mile since 1990, but over that time the number of passenger miles has quadrupled, so aviation emissions have risen steeply and continue to do so. The technologies for low-carbon long-distance flights are a long way off. Some marginal efficiency improvements are possible, but not enough to counter current projected growth in passenger numbers. Replacing fossil fuel is tricky because hydrogen is so bulky that it requires a completely new shape of plane, batteries are too heavy, and so-called 'sustainable aviation fuel' (SAF) is a misnomer because of the very high energy or land use requirement to create the fuel, and because SAF doesn't address the non-CO_2 effects. There is no getting around the need to reduce the number of flights. Only about one person in 10 flies in any given year, and 15% of those who *do* are thought to account for about 70% of flights.[63] So the reduction focus should be on cutting out unnecessary frequent flights. Flight 'offsets', by the way, are a misleading ploy, cynically promoted by many airlines because they know that if you or I wishfully buy into the concept, it will make us more likely to book a flight.

Shipping emissions account for around 2% of energy-related CO_2. We might expect the quantity of freight to reduce somewhat as the economy circularises. Key solutions for powering the remainder are green ammonia (that is, from renewable energy), hydrogen (compressed is bulky and liquified is inefficient, but both are options) and electricity for shorter distances. Luxury cruises, incidentally, are a climate nightmare.

For **land transport**, all cars need to be electric (both for climate and air quality), but they must be fewer and smaller, driven less and shared more. Electric two- and three-wheel vehicles are an efficient and practical urban solution, also improving congestion. Public land transport, especially trains, needs to be as good as it can get.

Even though some poor countries do need access to more energy, continuous efficiency improvements in almost everything we do and produce means that there is plenty of scope for cutting global energy demand. But we've already seen that energy efficiency improvements do not by default lead to cuts in total energy demand, and that to achieve this, a constraint is required. This takes us to the next action on the list.

Phase out fossil fuel via a **carbon price**. The important thing to re-emphasise here is that all the renewables and energy efficiencies that we could dream of won't on their own be enough to keep the fuel in the ground. Fossil fuel is generally so cheap to extract, so convenient to use, and our appetite for energy so limitless, that unless we specifically make its extraction expensive or illegal, production will continue at huge scale. The simplest mechanism to combat this is a high carbon price. The simplest point to impose it is at the point of extraction. The price needs to be high enough to cut the extraction rates. It is very hard to know how high this will have to be, but the way to know whether it is high enough is that we will see global fossil fuel use coming down fast.

In 2018, the IPCC non-committally estimated that a price of somewhere between US \$135 and \$5,500 per tonne of CO_2 in 2030, rising to between \$235 and \$13,000 per tonne CO_2 by 2050, should do it.[64] One explanation for why the world doesn't yet have carbon pricing is that the fossil fuel industry knows it will work, and their stranglehold over the process is sufficiently strong that, as things stand, policy-makers are more or less allowed to say and do whatever they like, so long as they won't succeed in denting extraction rates.

In terms of **renewable energy** generation, globally, electricity from solar panels is the biggest opportunity, with wind also very important and other sources such as tidal and hydro helping out a bit. There is also probably a contentious role for nuclear, but it isn't vast because for the foreseeable future it still looks too expensive and too slow to put in place, quite apart from the risks.[65] (Since the UK has the fifth worst sunlight per person of any country,[66] but an abundance of onshore and offshore wind, our clean energy mix will look more wind-centric than the global norm.) So far, the world's increasing renewable energy capacity has gone alongside rising fossil fuel use.[67] Remember that, in terms of our energy supply, the climate doesn't care how much clean energy we have. It only cares about how little fossil fuel we use.

> **Fossil Fuel = Total Energy − Renewables − Nuclear**

Having generated the electricity, we need to make sure it gets to everybody when they need it and where they need it. Most of it will be generated thousands of miles from the point of end use, so long-distance **transport** is required. A fairly small proportion can be put into the grid for immediate and fairly local use. The more efficient the electricity grid, the greater this proportion can be, as it makes it easier to balance fluctuations in demand if you can do so over a larger geographic region. For long distances, very high-voltage direct current is most efficient. An 800,000-V DC cable might lose just 2.5–4 per cent per 1,000 km.[68] This makes up to 10,000-km cables feasible in theory, even though the longest undersea cable, the North Sea Link between the UK and Norway, is currently only 720 km long.[69]

Nevertheless, a significant proportion of generated energy needs to be converted into a **storable form**, partly at the point of production and partly at the point of use, and at a scale that can't be met by batteries or by hydro storage, flywheels, compressed air, heat stores or any of the other mechanical methods. A key solution here is electrolysis to create hydrogen. Storing hydrogen is possible, both underground and in cylinders under pressure above ground. Unlike some of the mechanical storage options, hydrogen can also be used to manage both the daily and the seasonal differences between supply and demand.

Transporting hydrogen is also possible, but tricky. You can push it down a pipe at around twice the speed of natural gas, but it is so light that even at this speed the rate of energy flow is lower. And the molecules are so small that it is hard to stop them leaking out. Hydrogen can also be put on boats, but that is complicated, because even if you compress it to 400 times atmospheric pressure, it is still incredibly bulky – around three times the volume of oil per unit of energy. So, a completely new design of tanker is needed. Another possibility is to liquify hydrogen. The problem here is that you have to get it down to minus 253 °C. This takes a lot of energy in itself, and on board the tanker, it inevitably boils off all the time, so the knack is to use the boiled-off gas as fuel to power the boat. Once transported to the locality of end use, there will be further options for storage, and here batteries and hydro can also help, especially for short periods, but not at the scale that we will need it.[70]

The need for energy storage and transport is also minimised if demand is not only reduced, but if the timing is matched to supply by shifting energy-intensive jobs that aren't time-sensitive to times when there is a local surplus (contenders include some heavy industry, pumping water, charging cars and running washing machines).

A Sketch of the Sustainable Food and Land Transition (Items 5 to 8)

We need to eat fewer animal products and grow food more wisely, with fewer pesticides and fertilisers. We also need to rapidly deploy the emerging science of agroecology.

In a bit more detail, here is a quick guide on how to transition to a sustainable food system. What follows is a quick skim of 5 to 8 in my top-10 list. The sizes of the shapes in Figure 23 roughly represent how critical I think they are. On the left of the diagram are the things we can get onto right away, mostly without even needing much research and development, or infrastructure investment.

First up, the immediate priority is to reduce the amount of meat and dairy that we grow and eat. This will take enormous pressure off our land, making it easier to look after our biodiversity while providing enough food. It will take pressure off deforestation and dramatically reduce

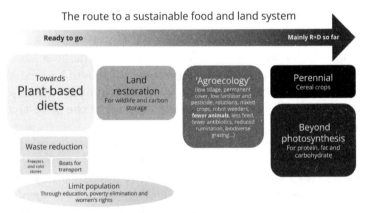

Figure 23 The size of the shapes gives some indication of their critical importance. The stuff on the left is ready to go right now. Moving to the right, there is increasing need for further research and development.

greenhouse gas emissions, especially methane (particularly critical for short-term climate mitigation, which we need to buy us time). Cutting out some of our most intensive animal farming will also mitigate disease threats, including the human risk from bird flu resulting from so many chickens at such close quarters and in contact with humans. Antibiotic resistance can partly be held at bay by reducing the *two-thirds* of all antibiotic use that goes to animals. Fundamentally dishonest elements in the meat industry are working hard to confuse the science lobbying to prevent the essential shift to more plant-based diets.[71]

The need to reduce waste in all stages of the supply chain is self-evident, though less critical than dietary change. In developed countries, this applies most importantly to cutting household and hospitality waste. Of less significance, but immediately implementable, is the elimination of air freight, partly enabled by *increased* use of frozen foods (which makes possible local fruit and veg out of season, as well as slower but less energy-intensive long-distance transportation).

Limiting population and reversing its growth will help to enable a sustainable food system, and as we'll soon see, the key to this does not require draconian policies.

Land restoration both for nature and for carbon sequestration and storage is totally essential. We largely understand how to go about it, and there is heartening progress in many parts of the UK and around the world.

More technical is the emerging art and science of *agroecology*: producing high-quality nutrition, at scale, in ways that restore soil quality after decades of abuse from copious fertilisers and pesticides. Agroecology involves such techniques as careful crop rotation, possible introduction of animals into the cycle, using nitrogen-fixing crops instead of fertilisers, avoiding use of ploughs (since this is terrible for the soil, habitat and its carbon storage), and the use of cover crops that ensure the land is never left bare (which leads to erosion).

Looking further to the future, the development of perennial crops in place of those that we currently have to sow each year will ensure that tilling is no longer necessary and that the soil is always covered.

All of this takes skill and hard work. The people who work our land need to be able to earn a living. Governments need to make this possible, and the power dynamics of the industry need to change so that big business interests don't squeeze out both small food-producers and less wealthy food-eaters.

Finally, we all need to get our heads around the idea of food produced without photosynthesis. In this way, renewable energy will be used to drive industrial processes to create macro-nutrients, protein, carbohydrates and fat. Why is this such a good idea? Because it is dramatically more efficient. Solar panels can capture at least 20% of the Sun's energy, and the process of turning that into human nutrition can be at least 20% efficient, too. That means that about 4% of the Sun's energy landing on a patch of ground can be turned into food. By comparison, photosynthesis typically captures about 1% of the Sun's energy that lands on chlorophyll. Yet a lot of sunlight will not manage even that, because the plant cover is not all year round, and even when the crop is on the field, some energy will slip through the gaps between leaves. The plant then takes that less-than-1% and uses it inefficiently to create plant matter, only a proportion of which is edible by us. So, the resulting efficiency difference between what is possible naturally and what we can do industrially is a couple of orders of magnitude. If the idea feels unattractive, there are two things to contemplate that might make you change your mind: firstly, the huge amount of land that can be liberated for nature, and secondly, the appalling conditions that animals are reared in today, the antibiotics they are fed on, and the fertilisers and pesticides that we pour all over our crops, many of which are then used as animal feeds. To say nothing of the reductions in carbon emissions.

The Circular Economy (Item 9)

The idea here is that we move away from a so-called linear economy in which stuff is taken out of the ground, made into things, used and then chucked away. Instead, the materials our things are made of go round in circles; they are made into things, and then remade into other things without ever being discarded. Circularisation is also enabled by making sure all the stuff we use is made up of sustainable materials, is built in such a way that it will last, can be lovingly maintained and repaired, and can be readily broken down into its constituent components when the end of its own life finally comes around, ready for the creation of something new. A 100 per cent circularised economy is an idealistic dream for the foreseeable future, but we can get much closer over the next decade or two, and that could be enough.

So, what will it look like? As citizens, we simply won't be buying so much new stuff. We might go to the shops just as much, but it will be to get things repaired or maintained, or to borrow something we don't need to own; to resell something we no longer need or to buy something second-hand. There will be opportunities for those who want to save money. More of our money will go on services and experiences and less on physical things, much less on *new* things and much, *much* less on junk.

At the retail end, our high streets will be refilled with preloved, repair and maintenance shops. Online shopping will be the same.

For research, development and manufacturing industries, the implications of transition to a circular economy are enormous. The opportunities are huge, because even though we will be spending less of our money on physical items, there will also be less spending by companies on raw materials. Those who keep their heads buried in the sand of the junk-ridden linear economy will face existential threats from reduced physical demand, and from the market that is moving away from their kinds of products made in their kinds of ways. We will be making things out of different materials so that they last and can be repaired. The clean energy transition will require much more of certain materials. Although we will need more of certain key minerals for a while, in general the mining industry will need to focus on producing less, but producing it better, and with fewer negative environmental and social impacts. The recycling industries will be growing both in volume and in the quality of their output so that, for example, recycled aluminium from cars can be reused for cars instead of just for beer cans. We can do a great deal to reduce the 25 million tonnes of plastic entering the ecosystem every year, but to solve the problem altogether, research is needed into bioplastics that don't last forever.

Taming the Population Curve (Item 10)

The rapid rise in toxins and resulting fertility loss may be about to take a few billion off our population whether we like it or not.

If that *doesn't* happen, the marvellous news is that if we just do some of the things we need to do anyway to make the world a better place, population will, of its own accord, go into a gentle decline. Nobody needs to be prevented from having kids. The critical thing is to make sure the decision to have children is a free and conscious choice, and neither undertaken nor avoided out of a sense of social or economic pressure.

The very good news from all around the world is that when a country becomes educated, when its people, especially its women, have rights over how they live, and when poverty is relieved to the extent that people are not fearful for their old age if they don't have kids, and can access proper healthcare, then population growth stops or even goes into decline. 'All' we need to do is make a better world for all of us to live in, and we will find that there will be fewer of us to enjoy it. Slightly ironic, but fantastic news from a sustainability perspective. The evidence from countries that improve the lives of their population is that when people make free and educated choices over whether to have children, they end up averaging less than two per couple, and the population falls.

The answer for those worrying about the scarcity of workers in an ageing population comes in three parts. Firstly, as the proportion of manual jobs declines, it is becoming possible for older people to play an increasing role in the workforce, and provided the jobs are meaningful and the work environment is positive, everyone benefits from this. The jobs don't have to be high-pressure or full-time, but being useful is good for one's mental health. Secondly, much of today's workforce is engaged in activities that are not essential for human quality of life, and a good many that are detrimental to it. Reducing these jobs liberates workers for essential roles. Thirdly, as we will see later, it is high time to revisit the rigid and out-of-date assumptions about the necessity of a constantly physically growing economy, behind which lie many of the clamours for population growth.

Why Isn't Population Decline a Silver Bullet?

That's the good news, but here's the bad news, and why population decline in itself is not a silver bullet. As I said in my previous book *There Is No Planet B*, 'One billion reckless people would easily trash the place.'

As living standards rise, so, by default, do carbon footprints. Just one European or North American has the same carbon impact as several hundred Malawians.[72] We have developed societies based on consumerism rather than citizenship; we think more in terms of what we can have than in terms of what we can contribute. So, when people amass wealth, they tend to spend it in ways that feed energy demand, resource extraction, pollution and destruction of nature. It does not need to be this way.

We urgently need a new spending model – one that leapfrogs the destructive lifestyles that are commonplace for the wealthy today and instead focuses on high-quality, sustainable living. With wealth comes responsibility. Being rich does not have to equate to a high-carbon lifestyle. Being over-rich does not have to lead to a space tourism hobby. We can ditch the shiny carbon-intensive status symbols for meaningful, low-carbon lives: lives based on human connection, mutual respect and high levels of equality. We can choose insulation, solar panels and heat pumps, instead of over-sized cars. We can all pay our fair share of taxes to enable an adequate welfare system, capable of providing universal basic services for all.[73]

This takes us back to a central point that whether there are 3 billion or 10 billion people on the planet, we *must* reduce our total energy demand.

We must consume less and be more.

How Do We Head Off Pandemics and Antimicrobial Resistance?

The changes to the food system outlined earlier will help a great deal. Resistance to every kind of antimicrobial antibiotics can be slowed by the reduction of quantities used on farms and by humans. The threat of diseases crossing from farm animals to humans will be hugely reduced by doing away with over-intensive animal farm systems and by reduction in farm animal numbers.

Diets can be improved partly by regulating the ultra-processed food industry, despite its powerful lobby. I am not sure yet in specific terms how the powerful pharmaceutical industry can be better incentivised to look after human health, but by posing the challenge, I hope it may bring the solutions slightly closer. And this is part of a wider discussion on the nature of business in the Anthropocene, which we will come to later.

Improvements in general health and thereby disease resistance should also result from a reduction of poverty, both in the developing world and in rich but deeply unequal societies, as well as from increased exercise, including through facilitating walking and cycling as part of daily life.

On the technical side, machine learning (AI) and RNA vaccines are rapidly improving our response to new disease threats, and progress is being made on new-generation antibiotics.

Finally, as with every other aspect of the Polycrisis, international cooperation is essential. For example, COVID-19 vaccination development could not have begun in January 2020 without China sharing the SARS-CoV-2 genome sequence, and international data on the spread of the disease was essential for enabling high-quality decisions on national responses.[74]

Overall, we won't cure all disease, nor eliminate pandemic risk, but we may be able to continually reduce the risks they pose instead of allowing them to rise.

Technological and Physical Solutions: In Conclusion

I've given a very quick sketch to illustrate, in just a few pages, how we can deal with the Polycrisis from a technological point of view. I've simplified things, and of course there are a lot of devils in the detail, but the central point is that the roadblock is nothing to do with a lack of technology, as an increasing number of scientists will attest.[75] Although it is important to keep pressing ahead with development of the technical solutions, it is also a form of avoidance to pretend – or wishful thinking to hope – that a technical focus alone will even begin to get us there. If that were the case, we would be getting there already instead of accelerating in the wrong direction.

While it is true to say that we are making technological progress on developing the solutions that we need, however well intentioned it might be, it is false, misleading and unhelpful to make the wider claim that overall humankind is making progress on the core components of the Polycrisis. The irrefutable evidence is that we are accelerating *into* the problems.

Making progress will not just entail deploying the right technologies at speed and at scale, but also refraining from over-use of the technologies that make things worse. This is going to be part of the new relationship that humans need to forge with technology. We will need to join up our solutions and find ways of deploying them that enable high quality of life while we are at it. All of this is entirely possible, but to understand why it isn't happening yet, we are now going to dig further under the surface.

4 THE MIDDLE LAYER OF THE POLYCRISIS

Politics · Media · Business · Inequality · Economics & Growth · Technology · Education

When we scrape off the outer layer to look underneath the presenting symptoms of the Polycrisis and their technical solutions, what do we find? It turns out that beneath the multitude of interconnected superficial problems, there are deeper issues that need sorting out, and that are preventing us from getting anywhere when we try to deal with the outer layer. This middle layer (Figure 24) will give us a first level of explanation as to why efforts to deal with the Polycrisis by focusing on its direct causes have been so frustratingly fruitless.

Let's look at what it will take for us get on top of the technically solvable problems that make up the outer layer. The difficulty for humanity right now is that the Anthropocene is a totally new context in which to be operating. Most of our social habits, customs and practices date back to the days in which we could get away with treating the world as a robust playground of a planet that could more or less bounce back from anything we did to it – or if we did smash a little bit of it up, there was always somewhere new to expand into. Now we find that almost everything about how we do life needs to be scrutinised by the question: **'Is this Anthropocene-fit?'**

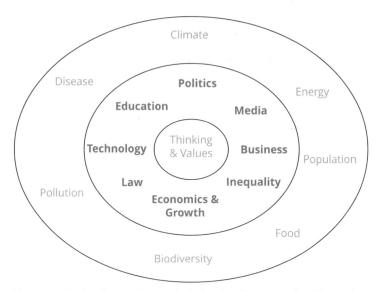

Figure 24 Under the surface of the Polycrisis lies a set of problems that are tempting to ignore but that are holding us back from progress. What politics, media, business, economics and education will we need? What relationship must humans learn to have with technology? And is progress possible without dealing with inequality?

Why Is It So Hard to Look under the Skin?

Techno-optimism is emerging as a new frontline of climate denial. What does it look like? Where does it come from? And why is it so harmful?

The central fallacies of climate techno-optimism are:
(1) 'We are making progress.' Supposedly evidenced by, for example, increasing renewable energy generation.
(2) 'Technology will drive the change. We will get there because the technology exists to do so.'
(3) 'The main changes that we need to see are technical.' In other words, the fundamentals of how we live, think, problem-solve, organise society, run the

4 The Middle Layer of the Polycrisis

economy and conduct our politics can remain largely unchanged.

A recent exchange on a well-known podcast epitomises techno-optimistic denial. With astonishing ignorance and naivety, for a man so clever in some other ways, Bill Gates blithely made the following statements on the popular podcast *The Rest Is Politics: Leading*, when Alastair Campbell asked him if he could reassure 'young people' (as if it was only them) who were worried about climate:

> *We are making progress . . . . The world does not end at two degrees. There's no stopping us passing two degrees . . . (but) in terms of your overall economy or livelihoods it's actually not a gigantic thing. Yes, you have to pay to make various changes. You have to have air conditioning . . . . The really bad stuff is if you let it go, say, above three degrees. Or if you live near the equator and you're dependent for your food on your yearly harvest.*[1] (Bill Gates, January 2024)

There are such huge problems with this narrative. You've probably spotted them all already. First of all, as we've seen, the statement that we are making progress is difficult to defend against the simple evidence of a carbon curve that is still heading upwards, representing acceleration into the problem. Whatever straws in the winds of 'progress' he thinks he sees, they have not yet translated into a foot coming off the accelerator, let alone being applied to a brake on carbon emissions. Second is the ignorance of, or failure to acknowledge, the seriousness of tipping point risks that could be kicking in *right now*, let alone at two degrees, or still worse three degrees. Third is the trivialisation of the impact on the many hundreds of millions of real people who, although Gates may never encounter them directly, do in fact live near the equator and rely on the yearly harvest. Fourth is his failure to take account of the possibility that so much suffering may leak into every part of the world through, for example, migration pressure, more widespread food insecurity and conflict.

One further way of understanding why it is that techno-centric solutions aren't enough is that they fail to heed some advice that is often, but probably wrongly, attributed to Einstein:[2]

> ***We cannot solve our problems with the same thinking that we used to create them.***

Tech-centricity assumes and hopes that our climate problems are only skin-deep, and that our fundamental approach to business, technology, politics and society can remain in place. It also comes from the fear of acknowledging the depth of the change that we need. It gains traction because it is comfortingly easy. It says, 'Just wait until solar power is cheaper than fossil fuel and the oil will stop coming out of the ground.' It is supported by the fossil fuel industry because they know that it isn't true and that the energy dynamics don't work that way. It gains ground because it allows political leaders to avoid having to grasp the nettle and explain difficult realities to voters. And it comes as a natural step on the Kübler-Ross Grief Transition Curve, which we will look at now to understand more about where techno-optimism comes from and why we need to get beyond it if we want to solve the problems we face.

The Kübler-Ross Grief Transition Curve, and Why Is It Relevant to Us Now?

Elizabeth Kübler-Ross was a psychiatrist whose simple model for understanding how people come to terms with their own impending death has now been adapted and evolved into a model for understanding how we react to difficult news of almost any type you can think of: from a death, to a redundancy, to an everyday inconvenience. We can use it now to understand a bit more about climate psychology and where techno-optimism fits in.

As we come to terms with grief or bad news, we tend to move along the curve in Figure 25 from left to right, although not always smoothly. Often we jump around. Sometimes we think we've moved on, only to find we have bounced back to an earlier stage. However, it's a useful way of recognising how we respond to bad news of almost any kind and any scale. Only when we get to the right-hand end of the curve and have accepted the reality of the situation, both rationally and emotionally, are we able to move forward and make the most of whatever that situation is.

A quick example. Suppose you are driving along the motorway and you run out of fuel. First up: 'This isn't happening.' Foot pumps up and down on the accelerator.

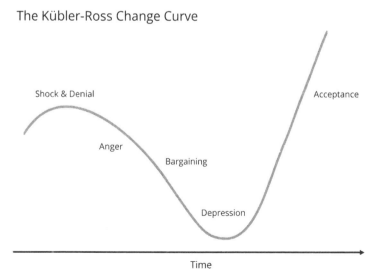

Figure 25 The Kübler-Ross Grief Transition Curve.

Pull into the hard shoulder and try the ignition a few times (Denial). Then you might swear a few times and thump the steering wheel (Anger). Next, some hedging around the problem, fishing for easy solutions; trying out whether you can get away with treating it as a more trivial problem: 'OK, I've got no fuel, but I'll use the whisky I've got in the back – or I'll hitch a ride ... or ...' (Bargaining). Next: head in hands, 'It's the end of the world' (Depression). Finally, and maybe not instantly, a realistic acceptance of the whole problem. As this emerges, you become, for the first time, able to do something useful to improve your situation. 'OK, I've run out of fuel. The engine won't run on whisky. And I have to stay with my car. I need to call a recovery service. I'm going to be very late, but I can let people know. I can do things to keep myself safe and warm while I wait. I haven't stopped rushing around for ages. It will be nice to chill out for a bit. I can do what I was going to do tonight, tomorrow night – which might work better in some ways if I make a few alterations to the plan ... .' Suddenly the problem-solving process might even begin to be enjoyable, and the solution, invisible at first, might in the best scenario even work out better than if we'd never had this problem to deal with in the first place. *Acceptance* is where we need to get to before we can start dealing properly with the situation.

In response to climate and the wider Polycrisis, society is spread out along the whole length of the Kübler-Ross curve. There is less flat-out denial than there used to be. More people acknowledge the problem in its many facets. Along the way, there are those who get angry when climate is mentioned, because it pisses on the chips of their otherwise more comfortable situations. Many, many more are bargaining with the problem. I see this all over the business world: wishful thinking that a 'net zero' strategy, with a few cheaply bought renewable energy certificates, is all that is required. I see farmers, restaurateurs, supermarkets who tell themselves stories that what is needed is not much change, or that someone else needs to move first. Some of the greenwashing I see is cynical, but much more of it is just bargaining with the problem and hoping it will agree to be smaller than it actually is. Climate techno-optimism falls into this category.

The bargainers, wishful greenwashers and techno-optimists are protecting themselves from the pain of looking the situation in the eye. This is understandable. But we all have to be stronger. We have reached a point where *nothing less* than full acceptance will do. People are scared of what they will see and what it will mean for them, their business models and their lifestyles. They are scared they won't see a way out. But increasingly, they are trapped, because they also know deep down there is more to look at, and this nags at them. Sometimes, and in some parts of my life, I realise I am still there myself. People in this situation need strength and courage, and are helped by reassurance. They need to trust that life on the other side of Kübler-Ross's depression dip is better than the life they lead right now, even if they can't yet see how. The psychological risks are real. There are some who, very understandably, get depressed by the Polycrisis and stay in that state because when they look at the enormity of it all they feel helpless. It's a nasty place to be. I don't think it is possible to stay there in good psychological health for long. You have to either move forwards to acceptance (and then action) or else retreat back to bargaining, or right back to flat-out denial.

> ***Some people see the world's lack of progress and jump to the conclusion that nothing can be done. It is essential to understand what you can meaningfully do about the situation once you see it. I hope this book will leave every reader at that point.***

> ### The Psychology of the Polycrisis at My Small World
>
> I work alongside some wonderfully smart people at Small World Consulting. The nature of our business means we spend a lot of time looking right into the eye of the Polycrisis. Our company ethos is that we do our best to see it like it is, to hide from nothing and to help our clients to do likewise. For this to work, everyone needs to have strategies for not just coping, but thriving. Most of my team seem pretty good at this. We tend to keep in good health, and the atmosphere is generally cheery. But every now and then someone has a difficult time for some reason or other, and I always find it hard to know whether or how much the Polycrisis is a contributing factor. We have occasionally had group sessions with a climate therapist who specialises in helping people stay healthy when they are working on this stuff. I like to think it is useful. Overall, I think life is richer and better for all of us when we confront reality, warts and all, especially when the alternative is going around only half alive in a fake world of inconsistencies that are patched over with happy clappy narratives, which we know deep down are just fanciful lies.

The Broken Trinity of Politics, Media and Business

These three components of the middle layer (Figure 26) need to be supporting each other and keeping each other on track. The media need to enable all of us to understand the issues and to know what is going on. Businesses, more than any other type of organisation, need to take the actions that directly benefit people and planet. Politicians need to provide the framework through which business is

4 The Middle Layer of the Polycrisis

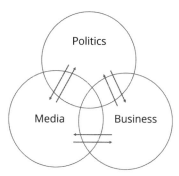

Figure 26 Politics, Media and Business need to support each other and hold each other to account in delivering what society needs, rather than bringing each other down.

steered to do so. The trinity of politics, media and business needs to be constantly collaborating; enabling and insisting that all three parts work for the collective good. Here in the Anthropocene, we need them, together, to be raising their game to a standard we have never seen before. They must make possible the high quality of understanding, of decision-making and of action that will allow humans to get on top of the Polycrisis – and thereby to thrive in the decades and, ideally, centuries to come.

Ironically, just as the challenges in the outer layer are accelerating in the wrong direction, so too the middle layer is collapsing; right when we need it to be reaching its highest ever standards.

Much of the media – both traditional and social – has been exerting its power to corrupt politicians in order to serve its own interests, while some politicians in turn have been working to undermine those parts of the traditional media that do the best job of upholding integrity and helping the public to know and understand what is going on. Business should exist to serve us, but the market economy, as it currently functions, has led to many industries making more profit the worse things get for people and planet: the defence industry makes more money the more conflict there is; the fossil fuel industry makes more money the more coal, oil and gas we burn; the entire energy industry makes more money the greater our perceived energy needs; most of the manufacturing industries are incentivised to have us over-consume; the pharmaceutical industry gets richer the more sick people there are; the ultra-processed food industry and the animal food industries are also incentivised against our best wellbeing. It doesn't stop with physical products. Most of the advertising industry makes its living persuading us to want and to buy things regardless of whether they will improve our lives or the ecosystem; the PR industry is similarly geared to manipulating our views regardless of the realities; the gambling industry makes up to half or perhaps more of its profits from problem gamblers – in other words, by destroying people's lives.[3] Cynical business interests have been pouring huge sums into both media and political influence. It's a mess.

There are many heroic good actors in politics, media and business, but so far they aren't winning the day. Why not? And what will it take to change the balance? That is where this book is heading.

When I started writing this book, in early 2023, from a UK perspective, it was clear that the current government was playing so fast and loose with the truth that whatever your traditional political instincts might have been, there was an urgent need to make sure they were not re-elected. In the time that I have been writing, things have moved on. Before the 2024 general election, the UK's Conservative Party had fallen to its knees, and in its attempts to revive its popularity seemed to have let go of the last pretences of a moral code; misleading ever-more frequently and blatantly and, beyond that, resorting to increasingly nasty tactics of stirring up hate and fear wherever it hoped this may save it some seats at the election.[4] In recent years and on both sides of the Atlantic, the fabric of our democracy has been hugely undermined.

In the UK, laws have been changed to criminalise peaceful protests.[5] Peaceful protesters have been convicted simply for letting the jury know why they took the actions they did,[6] and others have been prosecuted for letting the jury know their rights.[7] The once-proud BBC has been cynically infiltrated and cowed,[8] while Ofcom, the body tasked with upholding basic media standards, has somehow been influenced to apply radically less stringent standards to the ultra-right-wing, billionaire-funded GB News.[9] I never imagined it would be possible for the UK to sleepwalk its way out of democracy like this. How does this happen? On the other side of the Atlantic, things seem to be worse still. In fact, all around the world there are politicians who are well known to be routinely dishonest. Even outside of authoritarian regimes, many politicians are under the thumb of vested interests. Specifically on the climate emergency, at COP28 the fossil fuel lobby was joined in force by the meat lobby, to together ensure that the outcomes had no teeth.

We have to wake up fast: not just to undo the damage, but to find a new evolution in how politics, media and business can serve society.

The problems and blood-boiling anecdotes of how politics, media and business have been dragging each other into a downward spiral are so many that it has been hard to know which ones to pick. It is frustrating and enraging, but also depressing. At times I've felt the same hopelessness that I think many of us feel collectively; the hopelessness that explains how it comes to be that, in the UK for example, we allowed two million people to be unable to feed their kids properly,[10] a collapsing healthcare system,[11] fuel poverty,[12] and every piece of infrastructure and social fabric you can think of being run down for the sake of corrupt self-interests.[13] And of course, on top of all that, almost as a detail, an effective response to the outer layer of the Polycrisis is being held back in large part by cynical, dishonest tactics that are well funded by corporate vested interests.

I have written this book because it doesn't have to be like this. We can insist on something better, and by doing so start, at long last, to respond properly to the climate and wider ecological emergency. In 2024, the UK gave itself a chance to set a new course with a new government that got off to a good start in my view. But we must remember that the challenge is not just to reset standards but to raise the bar to a new level that will allow humanity to thrive in the context of the Anthropocene: our disproportionate and ever-increasing power over an ever-more fragile world. The reason honesty matters more than ever is that the consequences of dishonesty are now catastrophic.

Inequality

With or without the Polycrisis, I'd want to see the eradication of poverty and the opportunity for all people to have a decent quality of life. To enable that, we need a dramatic reduction in inequality: the concept of 'trickledown' economics is nonsense.[14] If I'm honest with myself and with you – which I am trying to be – I can't quite *prove* that it's impossible to deal with the Polycrisis without a fairer and more equal world, but it feels very likely to me. There are over eight billion of us *together* on this one Earth, like it or not. The fragility and interconnectedness of the system require ever-more cooperation. There is diminishing scope to make it work without everyone on the same side. All around the world, within and between countries, it is the poor who are the most vulnerable to any kind of disruption. They are the ones who live on the narrower margins and can't just buy their way out of trouble. However, even if the very rich don't care, it is risky at best for them to assume that they will be able to get away with ignoring the plight of the poor.

Can a Billionaire Be in Poverty?

In 2022, Oxfam reported that the 10 richest people had a combined wealth greater than the poorest 39 per cent of the world population.[15] In August 2024, these were the 10 wealthiest people in the world, with US $1.7 trillion between them (Figure 27).

For perspective, between them they have over 2,000 times more wealth than the utterly pitiful and embarrassing $700 million that the international community managed to raise at COP28 for the 'Loss and Damages' fund to compensate developing nations for the increasingly devastating climate impacts they are facing. Nobel Prize-winning economist Esther Duflo has called for a climate tax on global corporations and billionaires. She estimates that a wealth tax of just 2% on all of the world's 3,000 billionaires, along with raising international tax on multinational corporations from 15% to 20%, would raise a climate response fund of $400 billion per year.[16] This would be almost painless. And it is the kind of money that, if strategically targeted, could seriously help to create the change we need.

Humans find it easy to fixate on money, but the link between wellbeing and wealth breaks down completely after the first couple of million and can even go into reverse.[17] In fact, to me there is a degree of desperation in some of the antics of the over-rich when I read about space hobbies, and luxury yachts with solid gold taps. In Tibetan Buddhist mythology, Hell is a bad place to be, but not as bad as being a hungry ghost, who is never satiated and becomes ever-more hungry the more they receive.

\multicolumn{5}{	l	}{The Ten Wealthiest People in the World as of August 2024}
Rank	Name	Total net worth $	Country / Region	Industry
1	Elon Musk	$235B	United States	Technology
2	Jeff Bezos	$193B	United States	Technology
3	Mark Zuckerberg	$188B	United States	Technology
4	Bernard Arnault	$185B	France	Consumer
5	Bill Gates	$154B	United States	Technology
6	Larry Ellison	$148B	United States	Technology
7	Larry Page	$147B	United States	Technology
8	Steve Ballmer	$143B	United States	Technology
9	Sergey Brin	$139B	United States	Technology
10	Warren Buffett	$136B	United States	Diversified
Total		$1.7 Trillion		

Figure 27 The Bloomberg list of the 10 wealthiest people, 14 August 2024.[18]

Some of the over-rich (let's not call their wealth 'super') improve their public image and maybe assuage their consciences with philanthropy. However, I can't think of many examples in which this has been carefully and strategically directed at scale towards creating the systemic changes that global society needs. It is more often pushed towards superficial fixes with plenty of popular appeal that seem to me to be more like ego-massage oil. (This is particularly frustrating to me because for those looking to be more strategic, I am involved in some amazing projects where relatively small amounts of money can have huge leverage.) At its worst, philanthropy has a quietly corrupting influence.[19]

Thirty years ago, I was involved in an Oxfam conference exploring the meaning and causes of poverty.[20] The idea was to deepen the organisation's collective understanding, in order to better inform their response. The most memorable moment for me was when Frank Judd, a former director, stood up in a plenary to say, firstly, that he'd met a great many spiritually impoverished millionaires in his life, and secondly – and this is what really struck me – that unless Oxfam tackled *that* end of the poverty debate, it would always be doing first-aid. In other words, **unless we tackle spiritual poverty among the over-rich, we can never eradicate physical poverty.**

If the Shit Hits the Fan, Might the Over-Rich Get Away with It?

They might, but it is both a risky and a squalid strategy to pursue.

Perhaps one of the most chilling thoughts is that there are those who gamble on this. And maybe even groups who talk about this explicitly among themselves. We know there are over-rich people building bunkers and hide-outs in remote places[21] and developing plans for keeping their security guards loyal.[22] They might get away with it and be able to survive the planetary destruction that will take place when our growth mindset hits the buffers of the Polycrisis.[23]

So here is my call to the over-rich: think how depressing an existence this would be, especially in the knowledge of having contributed to the problem. Assuming you are not a psychopath or a sociopath, think of the background misery that will go with the deep inner understanding, as you sit in your bunker, that the rest of the world would have been better off if you had never existed. Alternatively, a few billion dollars of your wealth, strategically placed to work on the highest leverage areas such as those identified in this book, could work wonders in creating the systemic change we so badly need. You'd feel great, and, for what it's worth, the world would also love you. All you would really lose in return is space hobbies and gold taps.

Economics and Growth

It is unsurprising that an economic framework that dates back to the days when humans hadn't yet expanded into every corner of the Earth and weren't capable of wrecking the entire globe might need more than just a small tweak to be Anthropocene-fit. There are three central problems with our current economics:

- It is still predicated on perpetual physical growth, even though the context has changed from the days in which this was possible.[24] Humanity can no longer expand in physical terms and not expect disaster to unfold. This is because the Earth has been fully explored, and we have no foreseeable significant means of expanding beyond it. The logic of this is so simple and robust, yet it continues to be widely ignored and blanketed over. There are still precious few mainstream politicians who are prepared to challenge head-on the realisation that perpetual growth, in the traditional sense, is no longer feasible. Not all types of growth, however, are a bad thing. For example, it would be good to grow biodiversity instead of haemorrhaging it. And there is nothing wrong with growing human wellbeing, as long as we do so with an eye on the long term and on the other species that we share the planet with. We badly *do* need to grow global empathy. The most useful growth questions to ask are not 'How do we stop growth?' but 'What do we need to grow?' 'What is it OK to grow?' and 'What do we have to stop growing?'

- Linked to the growth problem, we chase the wrong metrics. Most notably, we are fixated with financial growth and in particular GDP, even though these correlate incredibly badly with the health and happiness of both the vast majority of people and the planet as a whole. In fact, Simon Kuznets, who developed the concept just 90 years ago, specifically warned against its use as a measure of welfare.[25] One way of increasing GDP, for example, is to create a society in which no one does anything to help anyone else, except as part of a financial transaction. GDP is so ridiculous that if you charge your elderly neighbour money to call in and check they are OK, and if your friends charge you to babysit your kids and if you charge them in return to help them with some DIY, then for all these things GDP goes up compared with a better society in which people do this stuff free of charge.[26] Incredibly, just as with the growth question, although the evidence condemning GDP as totally unfit for its role as the master metric of all economics could not be more clear or robust, as I write, there are precious few mainstream politicians with both the awareness and the courage to challenge its dominance. To give a further example, we use employment numbers as a metric, even though many jobs are more like bonded labour and/or are not of net benefit to people and planet, while others are both fulfilling and critically important to society. Some jobs contribute positively to employment metrics even though the world would be a better place if they didn't exist in their current form. Other valuable ways of spending time just aren't counted.

- A competitive market economy is incapable of dealing with global challenges that require global cooperation, unless it is accompanied by both cultural values and regulatory constraints to ensure that it acts in the common interest. Regulation alone will never be enough, because the complexity of looking after our fragile planet means that it will always be possible to exploit loopholes. So it needs to run alongside a culture in which all those who go against the *spirit* of that regulation find themselves paying a high price. A culture of caring for people and planet also isn't enough on its own, because a small proportion of the human population is made up of sociopaths – people who lack empathy and are more or less immune to shame (more on this later).[27] It will take both regulation and culture to contain them.
- If you look around the world today you can see what happens when sociopaths get into positions of great power in either politics, media or business. Just as with growth and GDP, the logic that a free market can't deal with global problems is simple and robust – and yet we still give airtime to free marketeers as if it is a legitimate ideology, support for which is simply a question of political preference. Support for unregulated markets is no more legitimate than support for walking off a cliff. So why do they get so much support? Because in a free market, wealth begets wealth and inequality grows.

Some people at the very top of the pyramid want that. Kate Raworth's widely acclaimed *Doughnut Economics* makes the case for an economy that finds a space between inner and outer boundaries (like a doughnut with a hole in the centre),[28] representing sufficient economic activity to enable wellbeing for all, but not so much that we exceed the limits of what the ecosystem can withstand. Rory Stewart and Alastair Campbell's popular *The Rest Is Politics* podcast makes, in my view, an important contribution to the media, providing accessible and intelligent analysis that is so badly needed. But when they interviewed Raworth on *The Rest Is Politics: Leading*,[29] while they superficially showed interest, in the end, for them, it was back to the 'real world' and they carried on almost as if she hadn't spoken. I greatly respect their political expertise, motivations and much of their analysis, even though I quite often, to use their own phrase, 'disagree agreeably' with them, just as they do with each other. But their treatment of Raworth is just one illustration of the frustrating difficulty in getting traction for the changes that we need to our economic framework. The challenge she laid down was so fundamental that it should have sent them right back to the drawing board until they had a proper response to it. So many otherwise smart people have enormous difficulty standing far enough back for the perspective that is needed. It is an incredibly rare skill to be able to think hard about the practicalities of running a country, while at the same time keeping an eye on a big-enough picture. We somehow need our political economists to be simultaneously astute enough to handle the here-and-now pressures of government, and imaginative enough to rethink the whole framework of operation.

Investment

If only we could get all the monetary trillions pushing in the right direction. The decision to own or manage assets of any kind is a decision to take responsibility for their part in the health of people and planet. Asset managers have a fiduciary duty in law to maximise value for their asset owners. What is critical is that 'value' be interpreted not just as financial gain (although there is still some place for that), but in a much wider sense, to include social and environmental value.

A small step, apparently in the right direction, has been the emergence of ESG (Environmental and Social Governance) as a buzz phrase for asset managers. It sounds good until you see the poor quality with which this is often assessed. Unless you have a very careful look at both the direct and indirect consequences of a company's activities, you can't see its real ESG credentials. In particular, if an asset manager buys a spreadsheet of company ESG scores from a ratings agency, they have probably wasted their money. A step forward from ESG is to focus on the positive 'impacts' of investments. Doing this properly requires deep consideration of the direct and indirect influence that an investment will have on enabling global transition to a sustainable future.

You don't have to be a big investor to be able to help a bit here. Make sure your employer knows that you want your pension scheme to get the basics right: divesting from fossil fuel and other ethical criteria. Let them know that you care about who they bank with. Switch your own bank. If you are wondering how to work out who and what to trust when it comes to ethical banking, Bank.Green lists fossil-fuel-free banking providers, searchable by country and by the kinds of service they provide.[30] It is sponsored by many reputable charities, and its top UK ratings align with my own assessments. And MotherTree runs a reputable service to help businesses – including my own – to switch their money.[31]

Law

Our legal framework and the way it works need to defend us against bad politicians, bad businesses and bad media. And a system that dates back to a time when the Earth could put itself back together again, however hard we kicked it, needs now to include protection for the planet – laws against ecocide: the destruction of the environment.

In the UK, Europe and the US, there has been an incredibly alarming undermining of democracy in recent years through clampdowns on peaceful protest, specifically relating to climate and to the Gaza conflict. Things have got so serious that a UN Special Rapporteur on Environmental Defenders has produced a damning report on state repression of environmental protest across Europe.[32] Here are some UK climate examples that should send a very nasty chill down the spine of anyone who values a free society.

- In April 2023, Judge Silas Reid ordered the arrest of Trudi Warner for contempt of court after she (legally) held up a placard outside a court informing the jury of their right to find a defendant not guilty if they felt conscience-bound to do so.[33] Conservative MP and solicitor general Michael Tomlinson took the decision to prosecute.[34] (Warner was eventually acquitted.[35])
- In February 2023, peaceful climate protesters were required by their judge (Silas Reid again) not to mention in court the reason for their actions, even though that reason was their legal defence.[36] Several people were handed jail sentences for contempt of court after mentioning the climate crisis during the trial.[37]
- In December 2023, Stephen Gingell was jailed for a massive six months just for walking slowly down the road for 30 minutes.[38]
- In April 2023, Marcus Decker and Morgan Trowland were given unprecedented and grossly disproportionate two-and-a-half and three-year prison sentences for unveiling a banner over a bridge.[39]
- In the summer of 2024, things escalated. Roger Hallam, one of the co-founders of Extinction Rebellion and a prominent figure in Just Stop Oil, was given a five-year jail sentence for his part in a Zoom call to plan a peaceful protest. Four-year terms were also imposed

on Daniel Shaw, Louise Lancaster, Lucia Whittaker De Abreu and Cressida Gethin. All these sentences were much harsher even than those given a few weeks later to rioters who had been throwing bricks, punching policemen and setting fire to hotels with people inside.[40] It is worth looking at this more closely. Specifically, Roger Hallam was convicted of conspiracy intentionally to cause a public nuisance, contrary to Section 78 of the Police, Crime, Sentencing and Courts Act 2022 (the very act that a UN Special Rapporteur described as 'extremely worrying'[41]) and Section 1 of the Criminal Law Act 1977. Evidence was provided in court that the protests saw 45 people climb up public highway gantries (risking their own safety and that of police and other emergency services) and led to an economic cost of at least £765,000, plus an estimated cost to the Metropolitan Police of £1.1 million. Further evidence asserted that the protests delayed more than 700,000 vehicles over four days of disruption and left the M25 'compromised' for more than 120 hours. Though laws were technically broken, the reasons for breaking them were not taken into account. Most of us, if we saw a baby struggling to breathe, locked in a car on a hot day, would break the car window to save the baby. Would it be fair that we were then sentenced for breaking the window? Roger Hallam argues he is doing the same – he is trying to save lives by putting the climate emergency on the agenda, especially for the UK government. The economic damage and disruption to daily life caused by these protests will be absolutely minuscule compared with the economic cost and loss of both life and livelihoods caused by climate change.[42] Noam Chomsky reminds us that every right we have has been fought for – none of them were given to us.[43] Women might never have received the vote but for the Suffragettes, and just as they are hailed as heroes now, climate protesters may well be held up as the heroes of our age in generations to come.

For all the bad news, there have also been a few heartening legal glimmers.

- In April 2024, the European Court of Human Rights set an important precedent by ruling that 2,000 Swiss women had had their human rights violated by their country's climate inaction.[44]
- In May 2023, the High Court of England and Wales sent the UK government back to the drawing board on its inadequate climate plan (for the second time), finding breach of duties in the Climate Change Act 2008 because the government had failed to consider the risks of policies and proposals not delivering emissions savings in full.[45]
- In August 2023, a judge in Montana ruled that the State's failure to consider climate change when approving fossil fuel projects was unconstitutional.[46]
- In November 2023, nine Extinction Rebellion protesters were found not guilty of smashing windows at HSBC's London headquarters, on the grounds that their actions were justified in the face of HSBC's £80 billion fossil fuel investments.[47]
- In December 2022, a former colleague of mine, on trial for non-violent direct climate action, was acquitted thanks to the heartening humanity of the jury. Judge Silas Reid (him again) had instructed them to find her guilty. However, she had managed, one way and another, to allow them to know the reason for her protest, and it took them just 37 minutes of deliberation to unanimously follow their consciences and find all the defendants not guilty. (In a very small way, I am happy to say that I may have helped, because the reference I had written for her, seen by the jury, described her as a climate consultant.)

- In February 2024, US climate scientist Michael Mann was awarded $1 million in damages for defamation after conservative writers denounced his work as 'fraudulent'. 'I hope this verdict sends a message that falsely attacking climate scientists is not protected speech,' Mann said.[48]
- In June 2024, the UK Supreme Court ruled that a local council should have considered the full climate impact of burning oil from new wells, overturning the previous absurd convention that only the relatively trivial emissions from the extraction process needed to be taken into account – a landmark decision that could put future UK oil and gas projects in question.[49]

These hope-giving instances are made possible by some wonderful environmental and human rights lawyers out there, working extremely hard and for nothing like the money they could earn elsewhere in their profession. ClientEarth, Friends of the Earth, Cornerstone Barristers and Hodge, Jones & Allen solicitors deserve name checks, but are not alone.

Justice is also badly perverted by the extent to which money talks in legal systems. Big corporations have much deeper pockets than environmental campaigners. On my doorstep, the legal case against the absurd new coal mine off the coast of Cumbria has, on the one side, the huge resources of West Cumbria Mining Ltd, whose lawyers are handsomely paid whatever the outcome of the planning and legal challenges because the legal fees are trivial compared with the potential mining revenues. On the other side are environmental lawyers who are paid thanks to crowdfunding but work at significantly reduced fees for a local climate charity, the trustees of which have had to ask some of their supporters to underwrite the risk so they don't have to sell their houses if they lose. Initial resolutions to grant planning permission for the mine were, from my observation, only approved because local planners were so fearful of the costs of defending their decision if challenged by the deep-pocketed mining company.[50] As I write this, Shell (2023 annual profits £22 billion[51]) is threatening to sue Greenpeace for what, for the latter, will be a crippling £8.6 million if it doesn't agree to cease forever its peaceful protests outside Shell's infrastructure.[52] For fear of litigation, Channel 4 TV shied away from broadcasting any of my scathing opinions in conversation with campaigning comedian Joe Lycett on the Shell-sponsored and, to my mind, bullshit-infused 'Future Earth' exhibition at the London Science Museum.[53] And as I've been writing this book, my editor has been understandably nervous every time I mention the names of bad companies and people – because if we are sued, even if we win, the costs of the case could be ruinous. As things stand, big money can use the law to bully all but the very brave and self-sacrificing into silence.

For businesses, regulation is necessary but will never be sufficient to ensure environmental stewardship. The issues are too complex and the world too fragile for legislation alone to be able to prevent carefree corporations from finding loopholes or making convoluted arguments that take years to go through the courts, or from just corrupting the process. The Cumbria coal mine is a good example again: West Cumbria Mining argued that the mine would be the world's first 'net zero' coal mine, but only by ignoring the emissions from burning the coal, relying on dodgy offsets and assuming a system to capture methane from the mining process that would be almost 100 per cent effective.[54] None of that is credible. All of it was supported by enough unprincipled legal arguments and questionable evidence from 'expert' consultants to clog up the planning and legal process for years unpicking the errors.

Unless the penalties for poor stewardship of the world are higher than the money to be made from going on the rampage, then an unscrupulous corporation can just put up with the fine and bag the rest of the profits. So, to repeat, for regulation to work, it needs to be backed up by a culture of enforcement; one in which everyone – not just a few activist campaigners, but staff, customers, investors and the wider public – screams when the spirit of environmental legislation is being perverted. Somehow, the price of disrespecting the planet needs to be higher than the perceived gains.

For the world of investment, as I've mentioned, a change in the interpretation of fiduciary duty is a must-have,[55] but is fiercely resisted by powerful lobbying. In 2022, 19 US state attorney generals signed a letter to BlackRock threatening legal action against them if they took decisions that sacrificed profit for the sake of the climate.[56] The letter sent waves of fear through the more responsible elements of the asset management community.[57]

The Relationship Between Humans and Technology

It is usual to talk about technology as if it is something that humans fundamentally control. We must do, because it is we who invent and develop it, right? But if we stand back – exercising that key skill that we need to get better at – and look a bit more carefully, it becomes a lot less clear that humans have the level of agency that we like to assume. In fact, here is how the dynamics currently work. A new technology emerges that offers an efficiency improvement. In a competitive economy with relatively few checks and balances, some company or some country adopts it and gains an advantage. Everyone else then has to adopt it or be left behind. Pretty soon the use of that technology becomes ubiquitous, regardless of whether its overall impact has been an improvement in the wellbeing of people and planet. It is only with huge effort that we have sometimes been able to curb some of the more obviously destructive technologies.

The most impressive examples of human agency over the broad trajectory of technology have been military. For nearly 80 years we have avoided using nuclear weapons, despite the efficiency improvement that they offer in combat. Similarly, we have mainly avoided the use of chemical and biological weapons. We manage this perhaps because the negative implications of their use are so emotionally triggering. Another widely trumpeted optimistic example has been ceasing use of chlorofluorocarbons (CFCs) in order to fix the hole in the ozone layer. And, aside from fossil fuel use, regulation is ensuring that the most polluting forms of transport and manufacturing are in decline in most of the world, with huge benefits to air quality and human health. These latter changes have been possible because no major industry has had to see its profits cut.

But on the other hand, if a technology offers someone a few billion dollars, it is pretty hard to hold it back. One critically important example right now, which will have sweeping implications for human existence over the coming years, is artificial intelligence (AI). At the moment, for example, if we see video footage of the wars in Ukraine and the Middle East, it still constitutes evidence of what is going on (well, mainly). But deep fakes are rapidly coming of age, and it looks likely that within a few years, most of the online footage not only will depict events that never happened but won't even have been created by humans. How then are we going to know what is happening in the world, or who and what to trust? AI can be used to expose falsehoods, but it can much more easily be used to create them.

Another danger is that AI algorithms are going to become better at learning how to make us think something than *we* are going to be at working out that this is what they are doing. To some degree, we can fend this off by developing our critical thinking and being extremely careful about what we allow to influence us, but in the end it looks a difficult battle to win. Turning very practically to the short term, wherever you live, AI is very likely to be deployed at the next election by at least some of the contenders, and in a more sophisticated way than it was the last time you voted. Whatever form of social media you use, it will be affecting the ads and the messages that are put in front of you, which will be tailored to an increasingly sophisticated model of how you personally think, based on every traceable preference and trait that you have ever hinted at through even the smallest online decisions. (This is jumping ahead, but please, do not take as evidence anything that you see or hear online unless you can independently verify the sources and their trustworthiness.)

Advocates of AI point out its many wonderful applications: pharmaceuticals and health, navigation and logistics, education, economics, entertainment and the ability to create almost anything you can think of more efficiently and perhaps better. However – and this is the show-stopper – it isn't sufficient to be able to say of AI, or any other technology, that it can be deployed for good purposes. You have to also be able to say that the good deployments will outweigh the nefarious uses. AI can be used to help both see and conceal the truth. It can be used to both create and cure disease. Its efficiency improvements can be used to both enhance and deplete quality of life. Very often, the destructive uses require less resource than the constructive ones – because it is easier to smash things up than to build or mend things, particularly in our fragile Anthropocene. So, for AI or any other technology, somehow we need a way of keeping the balance of application very firmly weighted towards the constructive end of the spectrum. The most worrying question for AI is 'If we wanted to steer its trajectory in any way, could we do so?' Some people talk about global regulation, to which my response is that the process needs to learn from the failure of the climate COPs, since we don't have three decades to get nowhere. In fact, increasing numbers of analysts believe we are on the cusp of losing control right now.[58]

The point about the need for humans to have a new relationship with technology does not apply only to AI. We need some technologies like crazy – to enable our transition to a sustainable future, for example – but the default trajectory of technology as a whole has so far been to just take us into more trouble. So, we have to learn how to be selective. It is a massive challenge in a market economy, and particularly one in which big businesses have huge power to drive their interests. The UK's Advanced Manufacturing Research Centre is one of my company's most exciting clients because of the way they have taken on this challenge. They have understood the need to think very carefully about the extent to which each project they get involved with helps to create the conditions under which the world can become more sustainable, and that efficiency improvements are not enough. They now select around £50 million per year of manufacturing research projects, using a tool we developed with them that is designed to get to the bottom of this question by asking deep and searching questions about the impact that the new technology stands to have on global society.

Education

This is not just a long-term challenge of ensuring that our kids have the skills and values they need for the twenty-first century. It is more immediate than that, and we are all in need of re-education for the Anthropocene. All of us need to develop, as best we can, the knowledge and skills required for the systemic change that we so urgently need. Of course there are technical skills but, relatively speaking, they are the easy bit. In order to thrive in the Anthropocene, humanity needs to get much better at thinking and behaving in the ways that will allow us to deal with the challenges that this new era presents. Later on, we will look a bit more at developing the inner core: the values, skills and other qualities that Anthropocene-fit humans need to have, which need to pervade our society, and upon the basis of which we need to be selecting our leaders. For now, let's just quickly say that central to the educational challenge are qualities such as honesty, empathy, respect, big-picture and joined-up thinking, critical thinking, self-awareness, resilience and courage.

No consideration of education would be complete without addressing the very unlevel playing field that our children are being educated on. In the UK and USA, like it or not, privately educated people are disproportionately represented in politics (one-third of UK Prime Ministers have come from a single private school). Whether or not one thinks such schools should be disbanded altogether, here are a few immediate comments for the UK's private schools. The traditional art of debating needs a complete rethink, because it is harmful to train kids to be able to win debates even if their arguments are bullshit (a term that I'll define later). Instead, the appropriate skills to instil include the ability to co-create understanding, and to change one's mind in the light of good evidence. The 16-year-old Boris Johnson wrote in his Eton school magazine, 'Strain every nerve, parents of Britain, to send your son to this educational establishment ... Exercise your freedom of choice because in this way you will imbue your son with the most important thing, a sense of his own importance.'[59] But a sense of importance needs to be replaced by an appropriate understanding of capabilities, including humility. The creation of a person who was capable of becoming chosen to be Prime Minister without possessing the honesty, empathy and humility to do the job well is not, in my view, something for a school to be proud of. Schools need to be teaching by example the skills of compassion and respect for all people, especially those from different backgrounds, yet many of those leaving the UK's most expensive private schools on their way to university, and then political careers, do so having never formed a friendship with anyone from a different social class and can often find their way right through the university system without ever leaving that bubble.

For universities, a coherent response to the Polycrisis spans the whole of the curriculum, the student experience, the research agenda and the ways in which they are funded. Just as in schools, students need to be taught how to think critically and independently, how to respectfully stand up for their beliefs, how to build relationships with people from different backgrounds and how to solve multi-faceted problems.

A Vision for the Middle Layer

I'm not going to pretend to have the perfect solution to the middle layer issues, but here is a sketch of the kind of thing that I think is feasible. I think what follows is highly attractive, and there are no *technical* barriers to putting it in place. I'm open to tweaking and debating the details, but I write the paragraphs below to illustrate that we could be doing so much better. What's not to like? Surely nothing should be holding us back from pursuing this vision, or something like it, at full speed? The barriers, as we will see, lie at the core, and we will get to those in the next chapter.

What might we aim for?
- **The political discourse is more honest, transparent and joined-up. If a politician is caught trying to mislead the public in any way at all, they know that it will be disastrous for their career, and also for any colleagues who stand by them. The rising mutual respect between the public and politicians is encouraging a higher quality of politician: people who are straightforward and kind, as well**

as dedicated to their work. So, the quality of political decision-making is becoming much higher, further enhanced by participatory democratic processes. These include Citizens' Assemblies and Citizens' Juries, in which representative cross-sections of society are given time, access to expertise, and facilitation to form opinions and report back on key questions. Such groups and processes are given teeth so that it is incumbent on governments to either act on their recommendations or produce more robust arguments as to why not. There is also rising coherence across our politics, supported by tools and mechanisms to enable joined-up thinking across the many dimensions of the Polycrisis.

- The media now also support and insist on honest and transparent government. The public selects its media as well as its politicians on the basis of honesty, transparency of motives, and freedom from interests and influences that conflict with the public interest. As a result, the Polycrisis and our increasingly coherent response to it features centrally in the news – because those outlets that don't do this are seen as irrelevant.
- The world of business is similarly becoming more honest and, with that, increasingly oriented to the interests of people and planet. We still have a market economy, but one which is infused not only with coherent regulation, but also with a culture in which customers, workers and investors all insist on properly looking after the environment and society, and have an increasingly sophisticated understanding of what this looks like. Through these mechanisms, together, perverse incentives in which industries benefit from poor outcomes such as conflict, ill health

and excessive energy demand are resolved. Companies that go the extra mile are rewarded while those that cut sustainability corners or greenwash are punished for it. Shareholders and asset managers of every kind, from pension fund owners to private and corporate investors, let companies know that their interpretation of fiduciary duty entails prioritising the health of people and planet over profits wherever these two come into conflict, and furthermore that they deem businesses to be negligent if they fail to be worthy custodians of the environment.
- We adopt an economic framework in which the core measures of success are closely and directly connected to the wellbeing of people and planet; politicians find they get yawned at if they fixate on GDP because we have all moved on from that rigid, outdated way of thinking.
- Wealth and income gaps between rich and poor are shrinking, such that higher proportions of the population are fully able to participate in society and have choice and real agency in their lives. The same goes for the basics, such as access to healthcare, healthy food, education, accommodation and transport. One of the many benefits of this is that population numbers have also started to decline earlier than expected, even though parenthood is more universally affordable.
- Ecocide laws exist and have teeth. There are growing numbers of cases of citizens winning damages from both corporations and states for climate impacts.
- The thrust of technology research and development is no longer driven by free market forces but by decisions over the public good. The market is steered both by appropriate regulation and by

- **cultural values that prioritise the interests of people and planet. Even AI is adequately regulated.**
- **In support of all of this, the educational curriculum encourages critical thinking, joined-up interdisciplinary thinking and global compassion, and the Polycrisis is a central theme in most subjects.**

None of the above is physically impossible. All of it would lead to a better outcome for humanity as a whole, as well as for all other species. For the overwhelming majority of people, there is nothing *not* to like in this broad vision. So now we are going to look at how to get there.

To answer what is *stopping* us from achieving these changes, we have to peel back the middle layer and look right into the core of the Polycrisis, and the heart of this book.

5 THE CORE OF THE POLYCRISIS

How we think and our values.

We've seen how there is a Polycrisis of technically solvable, urgent challenges, but also how so much of the way humans 'do life' is preventing us from enacting those technically available solutions: namely, the problems with our politics, our media, the way businesses and investments push us towards further environmental degradation, the hideous levels of inequality, our outdated economics, the inadequacies of our legal framework, our fundamental relationship with technology, and our education systems. We've seen how and why techno-optimism is part of the problem. This takes us to the core of the Polycrisis (Figure 28).

> **To solve the problems that we have created, we need to think in ways that are different from the mindsets that created those problems in the first place.**

We have to learn from what *hasn't* worked. I'd love to be able to learn from successes too, but sadly humanity has never yet successfully dealt with a Polycrisis like this, so there are no ready-made templates for how to do it. Some people point to fixing the hole in the ozone layer or the cleaning up of some cities, but these successes are really not of the same nature as even one dimension of the multi-dimensional Polycrisis. In fact, one of the key lessons from our failure so far is that it has been no good picking at the physical symptoms. We have to get to the *core* of the problem.

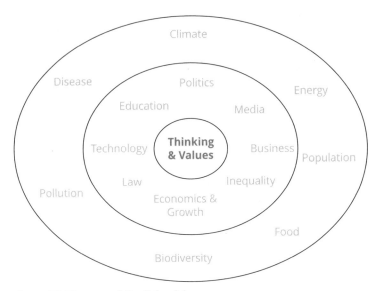

Figure 28 The core of the Polycrisis.

Those who work hardest on the middle layers can point to instances of progress – such as improvements in democracies and basic rights in some countries over the past century. But as we've seen, these middle layers are more commonly obstinately stuck – or maybe even going backwards – at a time when a huge and rapid evolution is now critically essential.

Luckily, it is possible for us to change our values and develop new ways of thinking, if we make the effort to do so. We *can* steer our evolution to enable us to survive and thrive through the Anthropocene challenge. So we are going to look now at the values and ways of thinking that we so urgently need to change in our politics, media and business, and very practically at what each of us can do to help bring that about. Just before doing so, however, it is important to acknowledge that the change we need is not all out there in the world around us. Some of it is internal.

So, before looking at what we need from the rest of the world, we are going to take a quick but important look at some of the inner qualities that we all need to develop – as best we can – in order that we can not only be the change we want to see but enable it as well. Since I don't claim to be any better at this stuff than anyone else, this section is also a note-to-self.

The Inner Challenge

Here are three simple models of characteristics we might need to develop if we are going to stand a chance of bringing about the practical changes in the world that we need to see. They range from quite a complicated list that emerged from a heavy-weight and sophisticated process involving many experts, to something so simple that it really belongs on the back of an envelope and would seem too obvious to mention if its message wasn't so important and yet so lacking in today's world. These models are not just about qualities we need as individuals, but about *collective* internal qualities: as Jonathan Rowson, co-founder of Perspectiva and former UK chess champion, puts it, this is also about our 'collective interiority'.[1] None of the models is perfect, so take your pick or use them to make up your own.

The 23 Inner Development Goals

These Inner Development Goals[2] (IDGs) came out of an impressive collaboration of universities, NGOs and businesses asking the question: 'How do we need to *be* if we are going to achieve the UN's Sustainable Development Goals (SDGs)?' In fact, these IDGs were developed specifically as a response to our failure to achieve the SDGs. Their aim has been to pull together a science-based understanding of 'inner development' and what is needed for a sustainable future. The idea is that we all need to develop these 23 capabilities, not just to become constructive cogs in a functioning society, but to offer some of the leadership that is required to get us there. The model is supported by events, resources and training programmes.

The IDGs (Figure 29) are largely self-explanatory, but if you would like more detail, see www.innerdevelopment goals.org. Note that many of the 23 qualities listed are ones that sociopaths, psychopaths and some of the world's most harmful politicians and business leaders have also honed to high standards – but, critically, such people also have yawning gaps. How many of us could score ourselves against all 23 capacities and be happy on every count? Not me. But the good news is that all of us have the *potential* to develop each one of them if we put our minds to it.

Just like the SDGs, the IDGs are not perfect. The value of both is that most of the world has agreed them, so they provide a common language, and they are good enough to work from. Comparing the IDGs with my own frameworks of thinking skills and values, which we'll look at next, there is reassuringly no contradiction, but the IDGs also call for courage, perseverance, communication and collaboration skills, as well as a 'relationship with self' which mine only hint at.

Seven Thinking Skills for Tackling the Polycrisis

Now for something simpler. Six years ago, I laid out a set of thinking skills that it seemed to me humanity needed to get much better at if we wanted to thrive in the Anthropocene. It isn't that these ways of thinking are brand new, it's just that they have become much more critical in our new context. There is no rocket science here. The framework originated from a scrap of paper, then appeared in my book *There Is No Planet B*. They have seemed to resonate with many people, so – with a quick apology to those who have already seen them – I repeat them here, in a slightly modified form (Figure 30).

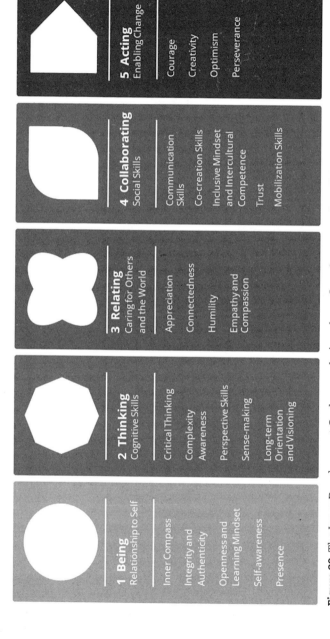

Figure 29 The Inner Development Goals were designed to reflect the personal qualities that each of us needs to develop in order to be able to help deliver the UN's Sustainable Development Goals.

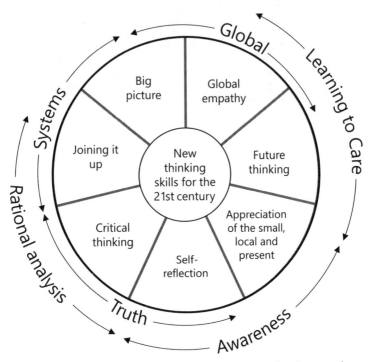

Figure 30 Seven key thinking skills that humans need to improve in order to deal with our challenging new context: the Anthropocene.

The 'Global' segments relate to the fact that humanity has globalised, like it or not. As a result, we need to be much better at seeing the **bigger picture** and **empathising** with all the other people on our communal planet. This is in all our interests, including our self-interests. The 'Systems' segments relate to the inescapable interconnectedness of the challenges we face. We need to get better at **joining up** our thinking, instead of trying to solve one problem at a time. Along with caring about all people wherever they are in the world, we need to improve our **future thinking**. That is because – more than ever before – the impacts of today's actions stretch further into the future. Two ways of thinking relate to inner awareness. The reason we need to get better at **appreciating small**, beautiful, local things is that the mindset of 'bigger is better' simply doesn't fit in a world that can't increase in size. **Self-reflection** and **critical thinking** are about our ability to distinguish fact from fiction. Self-reflection is needed not least because we need to understand our own place on the Kübler-Ross transition curve and our own emotional responses, in order that they do not blind us from reality. Critical thinking is crucial because we must have the clearest view that we can possibly get of the facts. We are in an environment in which the complexity of situations, coupled with dishonest or careless media and politicians, and the rise of deep-fake AI, make the task of ascertaining the facts harder than ever. We need to be as vigilant and astute as we can be. It is critical thinking which will get us there.

Three Essential Values

In 2019, I also wrote very simply about three core values that we can no longer live without if we are going to solve the types of challenges we face. I have been surprised at how often I am asked to focus on these when I talk to business audiences. I'll repeat them here (Figure 31) with a quick further apology for anyone who has read *There Is No Planet B*.

The need to radically improve our **respect for the environment** was laid out in the first part of this book. This has been articulated in great detail by a great many others and needs no further explanation just now.

The need to **respect all people** sounds to me almost too obvious to be worth mentioning, but you don't have to look very hard to see that we are not living this value. Within most countries, huge chunks of the population are treated as if they really don't matter much. The principle of equal respect for all people, as human beings, is frustratingly glossed over

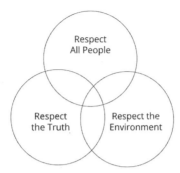

Figure 31 Three values that are no longer just nice things to aspire to; in the Anthropocene, humans won't be able to live without them.

by most world leaders, even though it could be such a powerful reference point for national and international affairs – and not least, as I write, the conflicts in Ukraine and Gaza. Instead, when heads of state routinely talk passionately about defending the lives of their own citizens, there is often a clear and profoundly unhelpful implication that the citizens of other countries matter less. The slogan 'America First' can be read as a selfish and greedy statement of disrespect for 96 per cent of the world's population.

Critically, the principle of universal respect applies whatever a person has *done*. Respect for a person does not imply endorsement or tolerance of a behaviour, or acceptance of a point of view. In the final analysis, this has to extend to every kind of criminal, political mischief maker, racist homophobe and corporate sociopath. People who harm others – whether through tax evasion, inciting wars, spreading misinformation or any other destructive behaviours – may need to be very firmly contained and sometimes punished (for the sake of deterrence and learning), but this can and should be done with respect. As Michelle Obama famously put it, 'When they go low, we go high.'

5 The Core of the Polycrisis

Respect for truth is going to be the focus of most of the rest of this book, because I think it is our point of maximum leverage over the outcome of the Polycrisis. It is both something we can't get anywhere without *and* something each one of us can change. On top of that, the time is right for a huge push on this, because even people who don't yet spend much time thinking about the environment are getting heartily fed up with some of the abusive dishonesty they have been enduring.

Turning specifically back to climate, though, when I look at all the worst decisions in the UK, they haven't boiled down to poor judgement, lack of information or incompetence. Fundamentally, they have come down to dishonesty. The debate over a new coal mine on the west coast of Cumbria, for example, can be spun as an argument with two legitimate sides to it, until you look carefully and find that one side is so utterly bogus that the protagonists must surely know it. Similarly, if a burger chain claims that soya protein isn't as good for human health as beef protein, a quick look into the science will tell them otherwise, and, surely, they have already had that look? Likewise, if an airline tells you your flight is 'net zero' because you ticked the 'carbon offset' box when booking and paid your fiver, they surely know it is nonsense.

Truth is so central that it drove the title of this book. We are going to dive into it now: what it looks like, why it matters so much, how to tell when you are getting it, and what we can all do to get more of it in our politics, media and business.

6 TRUTH – THE SINGLE MOST CRITICAL LEVER

Standing back again, remember that the context in which we now find ourselves requires us to make an evolutionary step change. To steer our way through the hazards of today's world, let alone those of the future, eight billion people need to cooperate as never before, to deal with systemic challenges of greater complexity than ever, and with less leeway for getting it wrong. The quality of the decision-making that humanity now requires simply isn't possible without radically higher standards of integrity than we see today, or perhaps have ever seen, from our politics, media and businesses.

Deceit, in all its forms, is a spanner that we can no longer afford to have in the works of our problem-solving.

So, we all need to sharpen our focus on this as one of the most critical leverage issues for anyone who cares about the future. The realistic possibility of raising standards here is one of my greatest sources of hope. We *can* get far more honesty into our culture. All we have to do is notice the difference and insist on better. If a politician says something they know is misleading, we need to treat it as abuse. The same goes for our media and for the businesses that serve us. I'm not just talking about straightforward lies, which are only the most visible part of the problem; I'm talking about all attempts to lead us to false understandings.

On climate, dishonesty has been blocking progress for decades. We now know that fossil fuel companies have deliberately muddied the waters on the science since the 1970s.[1] The complex nature of the issues has made it easy for them to spin their self-serving false narratives. None of us can possibly do all our own primary research, so we are reliant on being able to trust our sources; in particular, our media, politicians and businesses. In terms of policy, governments simply cannot begin to make sensible decisions without a foundation of honesty, free from vested interests. Yet in our politics today, false impressions are so easy to give and so hard to expose. When they *are* called out, there is usually so little price to be paid that the perpetrator has no regrets. The media, if they want to, can also feed us false impressions in a myriad of ways and face few consequences, judging correctly that for the most part the public will either trust them or not care enough to switch media. In the business world, I see an ever-more sophisticated approach to greenwashing, so far usually keeping just ahead of the public's ability to see through it. But we *can* change all this.

In all my work, I am always looking for the points of greatest leverage – the battles I can pick that will have the most impact. That quest has taken me to putting truth and honesty at the centre of this book. This is because in all the campaigns and debates I was involved in around climate, I came to see that dishonesty was wreaking havoc. I reflected more on the tobacco-industry-style tactics of the fossil fuel industry and its devastating effectiveness, still, in holding back the world's response to the climate emergency. I've seen slightly closer-up the deeply cynical and sophisticated falsehoods pouring out of fossil-fuel-funded so-called 'think tanks' in the UK. Similar tactics are employed in other industries, not least in aviation and more recently some elements of the meat farming lobby, distorting the media and confusing the public. However, it doesn't have to be climate and ecology that you care about most for honesty to be the most powerful and critical lever for you to pull. The same is the case if you are tearing your hair out about health and other public services, the long-term running down of infrastructure for short-term political gain, your ability to feed your kids and pay your energy bills, or if you simply feel that some of the big political decisions such as Brexit, or some of the wars your country has got into, were mis-sold to you. To start making headway on all these issues and more, a **climate of truth** across politics, media and business is what we most need. And the wonderful news is that *we can get it if we really want it*. We just have to not put up with anything less.

With higher standards of truth than we've ever known, we can actually start to make progress on the biggest, and hitherto most frustrating, depressing and infuriating issues of our time, as well as the most pressing issues for our daily lives. My hope is that by the end of this book you are going to have a very clear sense of agency; a sense of the simple but powerful things that all of us can do right now to push for higher standards of truth and in doing so, enable the change that has been eluding us on climate and everything else that matters.

To take the biggest-picture view, raising the standards of truthfulness is central to the urgent evolutionary challenge that humanity faces.

Pushing for honesty is perhaps the biggest thing you can do.

Why Is Deceit Easier Than Ever?

Because the issues are more complex than ever, so false narratives are easier to construct. Because media and especially *social* media can propagate enticing bullshit more easily than nuanced fact. And because we, the public, have been careless about this difference. To add to all this, the artform of deception has been getting ever-more sophisticated, with the proponents developing their expertise. The number of black belts in bullshit is skyrocketing. Lastly, as discussed, AI is already making things even trickier.

What Do I Even Mean by Truth, Honesty and Deceit?

I'm not going to get into a philosophical discussion here. That is a distraction, but a quick clarification of terms will be useful.

By being *truthful* I mean honouring the best view of reality that can be had with the evidence available (and yes, I am writing this book as if there is such a thing as a concrete reality). We can have different *views* of the truth – you may think a film is two hours long and I might think it is three hours long, but whatever I might believe or wish were true, if it was only two hours long then that is a fact; you are correct and I am wrong. There is no such thing as 'my truth', distinct from 'your truth'. If someone thinks the world is flat or that human-induced climate change is not a problem, or, like Bill Gates, that global heating doesn't get really serious until you get to 3 degrees of warming – that is not 'their truth', that is their false view of the truth.[2] Simply – they are wrong.

Honesty is a little bit more than truthfulness. Honesty means communicating in a way that is intended to enable others to get a clearer view of reality as you see it. It means you are helping others to understand something rather than to *mis*understand something. It would be dishonest and deceitful to say, suggest or do anything that encourages others to have a less accurate view of the world. It is deceitful for an advertisement to suggest, however implicitly, that a person will be happier if they do something or buy something that is against their interests. It is also deceitful to distract attention away from issues that are important to a person's full understanding of a situation, by focusing on things that are less important.

If only dishonesty was just about clear-cut lies. Instead, there are many forms of deception that are easier to disguise, at least as serious and destructive, and just as dishonest as simple falsehoods. Some politicians, along with advertising and PR professionals, have been working hard to study and refine the craft of deception for many years.

*Lies are only the tip of the iceberg, and just as the **Titanic** was sunk by the ice under the surface, so it is the more hidden forms of deception that are pulling us down into the Polycrisis.*

The many techniques in the deceiver's toolkit include: misdirecting our attention away from what matters most; omitting critical facts; implying that one thing leads to another when they know it doesn't; biased selection or biased assessment of evidence; subtle twists to facts including exaggerations and understatements; camouflaging difficult information by releasing it when there is more sensational news to grab the headlines, or burying it deep inside huge documents, or giving fake reasons for not releasing information (Figure 32). Perhaps the most prevalent form of deception is simply to allow others to do the dirty work by staying quiet in the knowledge that a colleague, a bogus think tank or a media outlet has been encouraging us to think something that they know not to be true. Sometimes it is difficult to prove the difference between a deliberate lie and an honest mistake, but failure to correct the record – and sufficiently loudly – often makes clear which one it was. Experienced liars can also be adept at deceiving themselves. There is speculation over whether Donald Trump even knows when he's lying. Others convince themselves that their deception is

Figure 32 The many forms of deception, of which lies are just the tip of the iceberg. The Titanic was sunk by the hidden ice under the waterline. (See also Appendix 1.)

justified – a so-called 'white lie'. Perhaps – and this is a generous interpretation – the most famously catastrophic example in recent UK history of a lie that was imagined to be white may have been Tony Blair's going with the 'beefed up' evidence for Saddam Hussein having weapons of mass destruction.

In Appendix 1 I have included a whole *Taxonomy of Deceit* in which all this is unpacked in more detail, complete with real-life examples, most of which are enough on their own to tell you all you'll ever need to know about the trustworthiness of a politician, a media outlet or a business.

Fortunately, while deceit is an artform that can be honed over time, so too is the ability to spot it, expose it and create real consequences for all those who were involved.

Blending Fact and Fiction: The Perils of Bullshit

It is impossible for someone to lie unless he thinks he knows the truth. Producing bullshit requires no such conviction.
Harry G. Frankfurt

Far more than a colloquial expression, 'bullshit' has a particular meaning. Stanford philosopher Harry Frankfurt wrote a paper and then a book – *On Bullshit* – defining and explaining its significance in our lives.[3] His definition is insightful. He describes bullshit as a blend of fact and fiction that has been concocted to persuade, without regard to which elements, if any, are true and which are false. The true stuff can be stirred in among the nonsense to make the whole cocktail more palatable and believable. **Skilful bullshit is a tasty mix of honey and poison.**

Long before becoming an MP, Boris Johnson was a serial propagator of bullshit.[4] Examples include reporting for the *Telegraph* that the EU wanted to standardise condom sizes because Italians had smaller penises. It *did* want to standardise for safety reasons, but the relative size of Italian penises had nothing to do with it. Other bullshit stories with grains of truth included his exposés on supposed EU plans to do away with prawn cocktail crisps and pink sausages. If you are tempted to think of this as a bit of harmless fun, note that Conrad Black, the owner of the *Telegraph*, wrote that Johnson had 'greatly influenced British opinion on this country's relations with Europe'.[5]

While Harry Frankfurt wrote that the bullshitter was unaware of the difference between fact and fiction, many of today's most accomplished bullshit merchants are keenly aware of that difference, and this enables them to use truth as camouflage. Once we hear a few verifiable facts, our trust goes up and our guard goes down. Skilful bullshit often looks lovely at a glance (Figure 33).

It is only if we take a discerning look that the various forms of deceit emerge from between the cracks, and the whole storyline falls apart, along with, we must learn to insist, the credibility of the perpetrator (Figure 34).

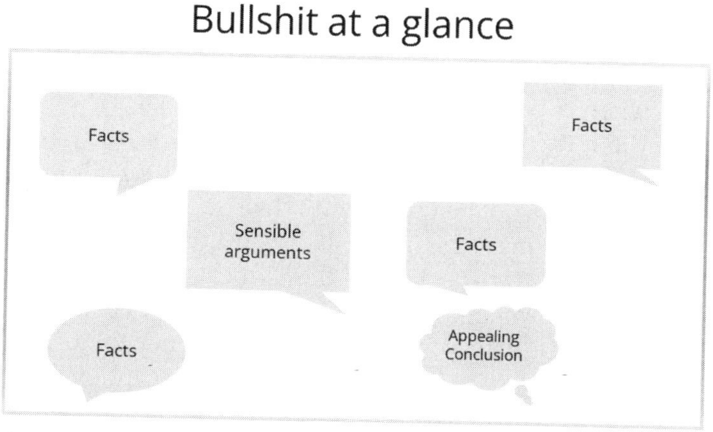

Bullshit at a glance

Facts   Facts

Sensible arguments   Facts

Facts   Appealing Conclusion

Figure 33 What bullshit looks like at first glance ...

Bullshit on close inspection

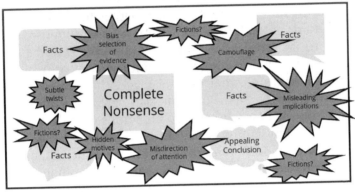

Figure 34 What bullshit actually looks like ...

Before leaving the subject of bullshit, I'll say that not all of it is entirely conscious. Sometimes the accomplished bullshit artist initiates an enchanting narrative that spreads through society. Some people propagate it with deep cynicism, but for the most part, wishful thinking is all it needs. The temptation to deceive ourselves can be as strong as the temptation to deceive others. In the world of business especially, plenty of bullshit greenwash boils down to weakness and lack of motivation to ask difficult questions about convenient storylines that hide awkward realities. In the same way, when a politician finds a convenient narrative that gives them the cover to pursue the line they want to take anyway, they may simply be too weak to ask tough questions. They may not be honest with themselves; or they may be cynically conscious. There is a continuum of bullshit consciousness (Figure 35). At the cynical end, reform is tricky. If the problem is self-delusion, maybe it is easier to fix, but the question of an underlying tendency to go with whatever sounds nice is serious. We are all susceptible. This is part of the reason why self-reflection matters so much.

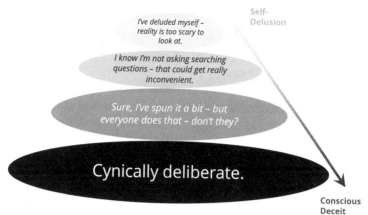

Figure 35 The bullshitter's spectrum of consciousness.

How Can Honesty Win the Day When Lies and Bullshit Are So Easy?

Every time a case of blatant truthlessness is uncovered, it tells us a great deal. A single, clear instance involving one person on one topic is like a thread that can be pulled.

A person's honesty is a generalisable quality in all their political work, their journalism and their business dealings. So, when one person has been provably dishonest about climate, 'levelling up', health or any other topic of substance, it should not be an incident that we have forgotten about in a week, a month or even a year. It is literally hundreds of times more significant than that (Figure 36). It tells us the following:

- If we find that they can't be trusted on one topic, then they can't be trusted on anything at all. We now know that they cannot be relied upon to put us accurately in the picture on any subject unless it happens to suit them, or they don't think they can get away with cheating us. It tells us that nothing they say constitutes evidence unless we can independently verify it. There is

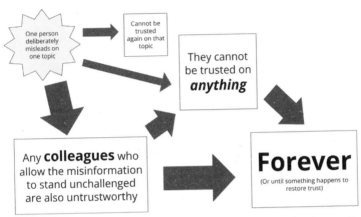

Figure 36 A single case of dishonesty in politics, media or business tells us so much. By remembering this and acting accordingly, we can create a climate of truth even though deceit is so easy and exposing it is so hard.

almost nothing we can learn from anything they say. It tells us they are unfit for political office, or to be a journalist or to run a business.

- It tells us that we can't trust them about anything at all in the future either, unless something quite unusual happens to give us convincing evidence of reform. People do not by default become more honest over time. So, a single incident early in someone's career can tell us all we need to know about their suitability for office decades later. I don't mean that redemption is impossible – but it is not the default. An apology and a few years to let the grass grow over our memories is not enough.
- Perhaps most devastatingly of all, it tells us that any colleagues who have knowingly stood by and allowed the dishonesty to go unchallenged are tarred with the same brush. They have failed to uphold an essential standard. If a politician has been prepared to tolerate dishonesty out of party loyalty, or for political or personal gain, or out of sheer weakness, then they, like the

original perpetrator, do not value truth sufficiently to be fit for office. If an editor turns a blind eye to a careless journalist, they are unfit for their role.

If we can treat evidenced deceit in this way, it becomes dangerous stuff for politicians, journalists, editors and businesses to dabble in. If we can establish a culture in which this is how truthlessness is treated, then we can get the standards we need, despite the ease with which deceit can be spread and the relative difficulty in exposing it.

For those readers who like a bit of mathematics, Figure 37 shows a simple equation (or strictly speaking an inequality) to show the conditions under which we can get that essential climate of truth. It also points to a few of the levers that can be pulled to bring about the conditions we need by working on the different factors within the equation. But if you don't like maths, all this is saying is that lies and bullshit are so easy to peddle that the price for doing so needs to be very, very high.

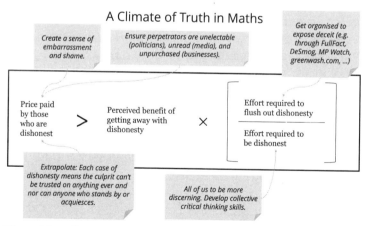

Figure 37 A climate of truth shown in mathematical terms. For deceit not to be worth it, the price to be paid has to be higher than the perceived benefit of getting away with it, by at least the ratio of the high effort it takes to flush out the deceit to the low effort of propagating it.

Why Do We Need Kindness Alongside Truth?

Because we are all in it together, like it or not.

The word kindness sometimes has fluffy connotations, but I'm using it to mean caring about other people and this entails being firm, tough and sometimes unpopular.

There are eight billion of us living interdependently in a globalised society on a now-fragile planet, and it is becoming increasingly easy for small numbers of disaffected people to smash it up. We will therefore deal with our Polycrisis collectively and collaboratively, or not at all. The idea of the over-rich being able to hunker down in their bunkers, while the rest of civilisation collapses, is as unrealistic as it is distasteful. So we had better learn to care about each other more than we do now, both within and between societies. This is not just because it is a nice thing to do, but because we can no longer get by without it.

Therefore, alongside honesty, the capacity to care about other people, whoever they are, whatever their background and however different their lives from our own, is something we all have to develop as best we can. These two qualities largely go together: why would you try to deceive someone if you have their best interests at heart? So, for example, if a politician isn't honest with you, it is a good indication that they don't care about you either.

If we want kindness from our politicians, our journalists and our business leaders, it is a very good idea to try to model it. Since this needs to be a universal principle, we need to treat people with respect even when they don't reciprocate. This does not mean letting them off the hook if they abuse us by trying to lead us up the garden path. It is easy to get angry, but holding people to account is not about revenge. It is about lifting the standard. 'We respect you as a human, but you are in the wrong job because you can't be trusted and don't seem to care enough about all of the other eight billion people in the world beyond yourself and your own tribe.'

Why Do We Need Strength Alongside Truth and Kindness?

It's often easier to let truth slip away, whereas it can take enormous strength to stand up for it, especially if you are up against your boss, a colleague or a friend. Also, if someone is pushing for the things you want – maybe Brexit, or a new coal mine in your neighbourhood, or promoting the company you work for – and they play fast and loose with the facts, it's so tempting just to let it go. What's the harm in allowing someone else to wrongly suggest that leaving the EU would liberate £350 million per week for the NHS, if leaving the EU is something you believe is the right thing to happen? The answer is that you open the door to a world of bullshit about everything that matters to you in the future.

For example, for Conservative politicians going into the UK's 2019 elections, even more strength was required. If a condition of your party allowing you to even stand is that you pledge support for a party leader who you know has told numerous lies – in fact, does it routinely and has even been sacked for it – then honouring the truth comes with a huge cost to your career options. In fact, the only option that leaves your integrity intact is not to support the leader, and to forgo your chances of being an MP at all. It takes a great deal of strength to call out your own tribe.

Right now, the cost of honouring truth in Europe and the US is rising. In Russia, China and many other countries, the price easily turns into years in prison or worse. As I write this – and irrespective of your own views on this issue – it takes strength to call out the state of Israel for bringing about the death of over 30,000 civilians, in response to the death of 1,200 Israelis on 7 October 2023, and for bringing starvation to a whole population, when you know this may have you spuriously labelled as an antisemite.[6]

My wish for people in the UK – and around the world – is to stand up for truth right now, while the cost of doing so remains relatively light. But it still requires strength.

What Part Do Psychopaths Play and How Can We Contain Them?

For the purposes of this book, I'm going to define psychopaths simply as people who don't have the ability to empathise with others but can be very socially skilled, charming and successful. They have little or no conscience. When someone else is happy or in pain, they don't experience any of the same emotion. Jon Ronson, author of *The Psychopath Test*, thinks that around 1% of the population has psychopathic tendencies, although because they are often successful in business life, the proportion among CEOs could be as high as 3 or 4%.[7] That's a lot. It means that you almost certainly know a few. They struggle to care about other people, even though they may have learned the skill of acting as if they do.

Many psychopaths live quietly selfish lives, but if society succeeds in holding the boundaries for them, they might still fit in without too much disruption. Just as in the population as a whole, a small proportion of them are very smart and ambitious. They can be extremely charismatic and learn how to say whatever will help them get where they want to go. Without the encumbrance of a conscience, they can have a competitive advantage over their peers. They are not derailed by shame because they don't experience it. There is no discomfort in being caught lying: it is just inconvenient.

So, smart, ambitious psychopaths can and do go a very long way in politics, in media and in business. The result is a trail of chaos and suffering. I'm not going to name names, but you might look around the world stage, and also within your own country, and ask yourself whether any of the big problems you can see might have been stirred up by psychopaths; or whether any of the politicians or business leaders you know might fall into this category. And you might ask whether our society currently has sufficient defence mechanisms to stop such people rising to the top. Or are we sometimes taken in by their charm, in sufficient numbers to elect them?

To keep them at bay, our skill at identifying them needs to be higher than their skills of disguise. And they learn all the time. The smartest ones have all the social skills you would dream of having yourself. In his book, Jon Ronson says that one test is to put a room full of people in front of a video of a horrific event – maybe babies having their heads crushed by tractors – then slam a door unexpectedly. Most people will jump out of their skin because the shocking film will have put them into a heightened state of alert. Only the psychopaths remain calm, because they have been watching the film with analytical interest but not with an emotional horror.

Turning to the everyday context of politics, we are looking for people who show no evidence of shame when they are exposed as having done things that would leave most of us wanting to hide in the corner. That is a huge advantage, because they don't give away signals that they have done anything wrong, and their opponents don't get the feedback that they have hit a nerve. There are people whose influence on the world has been chaotic and destructive, without remorse. Those in positions of power may, for example, have charmingly denied climate science for their own gain but with terrible implications for billions of others.

What can we do about them? We need to stop judging people by their charisma and look more carefully at their behaviours. Psychopaths have no reason to be kind or truthful, or even to look as if they are, except to the extent that it suits their purpose. Over time, they leave a trail of dishonesty and disruption to the lives of others – probably personally and at scale. We need to get better at spotting it and not putting up with it. We can't afford the spanner that they throw into the works, so we need to learn how to contain them. We need to be less gullible.

Charisma, charm and wit are not reasons to trust someone or to give them your vote.

Remember, when listening to psychopaths, that everything they say is simply designed to get them the result they are looking for. So, you cannot use their words to inform yourself about any aspect of the world. What they say can only be used – and even then with caution – to try to understand what it is that they are trying to achieve. Over the next three chapters, we will look in a bit more detail at how we can raise standards of truth in the critical areas of politics, media and business, and part of the reason for this is to help contain our psychopaths.

7 GETTING TRUTH INTO POLITICS

I'm going to be talking about truth across a range of areas, but, as already mentioned, politics is one area where the issue is most pressing. In this chapter I'm going to specifically reference the UK – since that's where I live, and what I think I know best – but so much of this applies to all countries that call themselves democratic. What follows is not a recipe for perfection, and I won't try to bottom out every detail, but it looks clear that relatively simple, practical improvements stand to have a huge impact.

Some people say there has always been lying in politics: that it's just the way the world works. The first thing to notice about this argument is that it is often put forward by the most dishonest politicians because it normalises and justifies their own approach. They would love us to believe that dishonest people are still worth voting for. The second observation is that, under Trump and the 2010–24 UK Conservative government (although some of the rot may have started before their time), we have experienced an eruption of deceit unlike anything I've known in my lifetime. It may not be running rife across the whole of the Earth's surface, but it has certainly been raging on both sides of the Atlantic, despite the Biden administration's short-lived attempts to bring it somewhat under better control. A decade or two ago, in the UK and the US, politicians sometimes misled, but if you were caught, you were likely to be in a lot more trouble. The third and deeper observation is this:

Whether or not we've always tolerated lies and bullshit, here in the Anthropocene we can't afford to.

Yes, honesty is needed from all of us, but it is the honesty of those powerful few who influence the lives of millions or even billions, whose choices trickle down through us as their captive audience, that has the most seismic and far-reaching effects. In 1994, the UK Prime Minister John Major tried to raise standards by setting up the Committee on Standards in Public Life, leading to the seven so-called Nolan Principles,[1] lack of adherence to which has been widely lamented by many commentators on the recent parliamentary ethical collapse.[2]

The Seven Nolan Principles of Public Life

The Nolan Principles came about in response to a series of scandals that had left the public disgusted, but would today seem depressingly normal. The committee, chaired initially by Lord Nolan, came up with seven principles (Figure 38). While they were specifically developed for those in public life, you can argue that they are just as important for everyone whose job affects the public – and that, of course, is everyone who has a role in society; every citizen, including anyone connected with the media or involved in business.

Selflessness is interesting to me because I think it goes too far. One hundred per cent selflessness is more than can be expected or is required and is even a bit illogical – it doesn't work to have people looking out for others but not looking after themselves. I think it is legitimate for all of us to take a proportionate interest in our own wellbeing, and taking this principle as described does not actually allow for that. This matters because by overstating the need not to be selfish, the credibility of this huge principle – and the other six – is slightly undermined. Some people have made the case that the vast majority of politicians have great **integrity** and that we have been let down by a high-profile few. I would love to believe that. My Taxonomy of Deceit (see Appendix 1) itemises some of the ways in which **objectivity** is breached, and to me this also breaches the **honesty** principle, which I need to emphasise: that truthfulness needs to mean so much more than just not telling lies. **Accountability** means taking responsibility for things that happen on your watch. So if hundreds of people had their lives ruined because of a faulty computer system and a widespread cover-up in your organisation, the buck stops with you. **Openness** should mean, amongst other things, that the reason you advocate for something is the real reason you want it to happen, rather than the one you think you can most easily sell to your audience. **Leadership** as defined here was critically missing, for example, in the 2024 UK Conservative Party in which every single MP has tolerated and quietly allowed routine dishonesty in their own party.

> *It is not OK to put up with the low standards we see in public life today.*

If we turn off certain bits of our brains, it can be fun – at least in the short term – to go along with a pack of lies and misinformation. A witty and charismatic person can lead us into a wonderfully enchanting fantasy world that takes us away, for a little while, from the problems of the moment. It can even feel good to allow our frustrations with life to be stoked up into anger and misdirected towards any group that can easily be blamed. I'm not

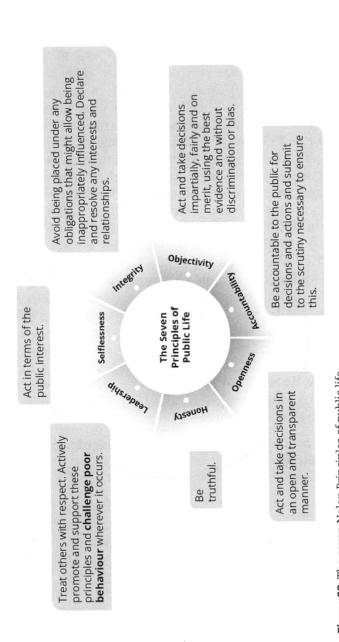

Figure 38 The seven Nolan Principles of public life.

saying many of us do this deliberately, but collectively we haven't fought hard enough to hold onto that boring inconvenient thing called truth when the alternative has been so seductive, temporarily shielding us from that lowest point of depression on the Kübler-Ross curve.

In the end, however, we come up against the consequences. We find that the charming, entertaining people, sometimes with interesting haircuts, who felt so good to vote for at the time, turn out to have been taking us, our family and our loved ones for a total ride. The fabric of society and essential services are breaking down around us, there is crushing inequality, and our beautiful planet is careering into a climate and ecological breakdown that threatens everything we care about.

To give an example, in July 2023, Prime Minister Rishi Sunak and Grant Shapps, then Secretary of State for Energy Security and Net Zero, both justified expansion of UK oil and gas production in the North Sea, against the clear advice of the IPCC, the UK's own Climate Change Committee (CCC) *and* the scientific community as a whole. They claimed that this was necessary in order to avoid household energy prices rising sharply. However, they both would have been well aware that UK oil and gas prices reflect the global energy market and are barely influenced by whatever the UK does in the North Sea. In order not to have understood this, they would have needed to be spectacularly incompetent in their roles. To me, it is far more plausible that these statements were cynically dishonest. But they were barely challenged, because in the UK, that kind of misinformation had been normalised.

While in this book it is the Polycrisis that has led us to focus on honesty, it doesn't really matter what it is that you start off being most bothered about. Whether your initial concern is for the health service, or social justice, or basic access to food and energy, or 'levelling up', or your loved ones surviving a pandemic, or the question of whether the

UK should be in the EU, or which wars, if any, we should get involved in and how, it is clear that we need the best possible view of the evidence. Without that, it is not possible to make adequate decisions.

Anyone who plays fast and loose with reality needs to be ejected from positions of influence.

Ironically, just as we need honesty more than ever before in our society, in recent years we have been getting less of it than ever, on both sides of the Atlantic and elsewhere. In the UK, how, for example, did we even come to be in a situation in which a man who was sacked for dishonesty as a journalist was selected by a major political party to be a parliamentary candidate? Alarm bells should have been screaming at that point. How did a person with such a dishonest track record then become an MP? How was he *then* able to get away with peddling misinformation after misinformation?[3] Why did every single MP not complain about the '£350 million a week for the NHS' claim on the side of the Brexit campaign bus, regardless of their views on Brexit? How did that not lead to a career-stopping loss of credibility? How did the perpetrator get to be elected party leader and then Prime Minister? He was fun for many, and his misleading narratives were enticing. He was telling a disenchanted electorate what it wanted to believe: that they could save money, have more freedoms, have more power, and most of all, that those things were theirs by right. It was seductive. But dishonesty is also abusive.

In the spring of 2024, dishonesty in UK politics had become part of everyday life for many ministers, and allowing it to go unchallenged was routine for every one of their colleagues. On the other side of the Atlantic, crimes against truth have been even more extreme, and globally there are plenty of worse examples still. So-called 'populism' has been erupting all over the world.

Is Dishonesty New in Politics?

It is not a new phenomenon. There has always been some dishonesty in political life and the least honest politicians love to point this out, because it normalises their own behaviour. However, things have got worse. A lot worse. I don't have definitive knowledge of where the rot started, but two events stand out. The first is Margaret Thatcher, trailing in the polls in 1981, doing a deal, at a secret and long-denied meeting at Chequers, with Rupert Murdoch to support his bid to control nearly 40 per cent of the British press in return for News International's support of her.[4] In doing so, she began a major unravelling of the independence of the UK press. The second is Tony Blair, a successful Prime Minister by some measures, but one who reportedly told many lies, from fabricating his CV,[5] right through to misleading the British public over the strength of evidence for Saddam Hussein's weapons of mass destruction. Whether or not he honestly, but mistakenly, believed this latter most serious act of misinformation to have been a 'white lie' taken in the global and national interest, one of the consequences has been to help to normalise political untruths. How he became a role model for political lying was carefully catalogued in Peter Oborne's 2005 book *The Rise of Political Lying*. Later, Rory Stewart's account of his time as an MP under David Cameron, and then as a cabinet minister and prime ministerial candidate, reads as a catalogue of deceit from his colleagues: to the public, to him and to each other.[6] It is clear to me that since then, under Boris Johnson, then Liz Truss and Rishi Sunak, things collapsed further still. Much of this is again carefully, chillingly and depressingly catalogued by Peter Oborne in a further book, *The Assault on Truth*, and his website political-lies.co.uk – both highly recommended.[7]

I have used UK examples here, but if you live elsewhere, I am sure you can think of similar examples of dishonesty closer to your own home.

Why Do the Wrong People So Often End Up as the Most Powerful Politicians?

In her book *Why We Get the Wrong Politicians*, Isabel Hardman describes the process of becoming an MP in the UK – the need for personal finances, the requirement to kowtow to the party at both the national and local level, the brutality of the process and of the job once you get it.[8] It is a system that simply does not lend itself to the people who would be best at making the nation's biggest decisions finding their way into those roles. In the US, the financial resources required to run are even higher, deselecting all but a tiny proportion of potential candidates before the race even begins. On top of all this, in the UK, the first-past-the-post electoral system ensures that the candidates and parties that win are those that have the most empty political space around them, whereas those who compete against others around the most commonly held (and sometimes the best) ideas all become unelectable by splitting each other's votes. All of this is so obviously inadequate, yet so persistently hard to put right because of people and parties looking after their own interests rather than those of the greater good. Finally, the life of an MP – or at least the life of a morally conscious one – involves verbal abuse, physical threats, hideously long and antisocial hours, and being torn to shreds for displaying some of the essential traits of an intelligent, mature decision-maker, such as seeing two sides of an argument or changing their minds in the face of new evidence, or even personal reflection. It can even involve being murdered.[9]

The media could help with many of these problems, but most of them don't. While their most central job needs to be to hold power to account, the world's traditional media often have a clear track record of carelessness with the truth and pushing the agendas of their billionaire owners. And yet people still buy and subscribe to them without asking careful questions about trustworthiness, let alone respect. Meanwhile, social media increasingly offer an influential platform for fake narratives.

Are the Results of the Post-Truth Experiment Coming In?

One way of taking heart from this epidemic of post-truth is that it can be seen as humanity running an experiment to test whether that boring old honesty stuff is really as important as some people have been making out. In the UK at least, the results of that experiment are now very clear: our society cannot thrive for long if we are careless with the concept of truth. The good news is that if standards of honesty can decline, it means they are variable. So it suggests they can also get better.

I hope that in the UK at least, the pendulum could now be starting to swing the other way. Even in 2021, a detailed survey of 6,500 UK citizens found that honesty was the most highly valued characteristic in a politician; far more valued than being clever, or hard-working, or getting things done.[10] The 2024 UK general election result might turn out to have been a critical moment, marking the start of a major and long-term rejection of post-truth: a rejection of routine dishonesty from the people we entrusted with our votes. Hearteningly, as I write these words, a wave of UK riots, revved up by fake narratives on X (formerly Twitter) and other social media, have been countered and quashed, at least in part by a much larger wave of peaceful counter-protest that looks to have taken many of the rioters out of their insular social media thought-bubbles and confronted them with the reality that most people are disgusted by what they are doing, and has even presented the idea that it might be more normal to want to treat refugees with respect. The background narrative behind these riots – that immigration rather than rising inequality and falling opportunities was to blame for the rioters' sense of not being valued – has been stoked and promoted by many of our worst media outlets and politicians.

If We Let Things Slip Further, Where Does It Lead To?

A very, very, very bad place.

I live in the UK. We like to talk about ourselves as a robust and unshakable democracy. The reality has never been quite as good as the national rhetoric has suggested, but compared with many places it has been wonderful throughout most of my life. I never thought I'd see it deteriorate the way it collapsed in the decade leading up to the July 2024 UK election.[11] In its final weeks, Rishi Sunak's government seemed to be pouring deceit all over the electorate in a constant flow.[12] So much so that to me it even looked like a deliberate policy. We had collectively got used to it. Increasingly, it takes effort to stand back and remind ourselves quite how disgusting the everyday behaviours we see from our government actually are. When the truth breaks down, anything goes. Absolutely anything. The war in Ukraine is justified by lies. So was the Second World War, along with the associated Holocaust. So are all the world's present-day genocides. I can't think of any examples of truth-defying regimes spontaneously U-turning; deciding that enough is enough and that they need to uphold standards of some kind or another. The boundaries are held either by other countries or by their own people. In the UK and the US, it is incredibly dangerous *not* to imagine how easily things will get truly terrible if we don't stand up to it now.

Five Criteria for Assessing a Politician's Honesty

Now to get practical. Here are five simple criteria for making up your mind about a politician's integrity (Figure 39). They aren't perfect, or complete, but I'm suggesting them as a simple framework. Please feel more than free to tweak them to take into account anything important that you think is missing. At the end of the day, I just want you to have the best possible means of working out who you can trust, so if you end up with something better than my suggestion, that's great, and please let me know about it. I want all of us to have thought carefully about how we decide who to trust, and whether our method is good enough.

Figure 39 Five criteria for assessing a politician's honesty.

Just to emphasise yet again: this framework isn't party-political. So, when you are using it, if you find that all the politicians in one party or another are coming out badly, then you can take that as evidence that there are more general problems with standards of honesty in that party. When it comes to voting, you might have to vote for the least-untrustworthy option that stands a chance of getting elected. That's OK. If everyone does that, standards will rapidly rise.

(1) **Honesty track record.** To what extent has this person been consistently careful to impart the best attainable view of reality as they see it? Or have there been one or more indisputable examples of them saying something in relation to a policy issue that they know to be incorrect? Have they said things that they know are misleading?

(2) **Transparency of funding, interests and influences.** Are their sources of funding, both personally and politically, made transparent? Tortoise Media has a website documenting how UK MPs are funded, for example.[13] If you don't have a similar website in your own country, please campaign for one. A short time browsing through it will give insights into corrupting influences.

(3) **Freedom from corrosive interests.** Being transparent is an essential step, but it isn't enough. We also need to know that the funding and lobbying they receive doesn't stand to influence their political decisions against the public interest. And we need to know that the people they meet and those who advise them also have high standards of honesty.

(4) **Scrutiny of information sources.** Do they select the highest-quality and most robust evidence, or do they pick their sources to give the result they want to see? Do they have the critical thinking awareness to understand the biases that may lie behind the information sources they receive?

(5) **Expecting honesty from others.** This fifth criterion is about whether they honour their responsibility to help uphold standards in public life. Are they happy to stand by while others are dishonest, or do they make a stand for honesty even if it is unpopular, or even if it means calling out someone in their own party? By having the expectation that our politicians uphold high standards in others, we create an environment in which our political figures know that if they are careless with the truth they will become isolated. They will become a reputational risk to all those around them.

What Is Not Included in the Five Criteria?

I've stayed clear of politician's personal lives, even though there are times when a person's conduct might shed light on their integrity more generally. I've stayed out of it because we need to cut politicians some slack and give them some privacy. Delving into private lives usually ends up being more about mudslinging and humiliation than about helping us to select the best political representatives. I also haven't included any criteria on coercion or bullying: the pressure that MPs sometimes put on each other to say and do things against their better judgement or consciences. I have not included a politician's voting track record, because this does not specifically relate to their honesty. In the UK, for example, there are times under the whip system when MPs are legitimately asked to vote against their personal preference for the sake of coordinated government. However, if they then defend their vote with arguments that they do not believe, that is a different matter.

I have also not included some of the other characteristics that I think are essential for good politicians. The most fundamental omission for me is kindness. I think we should be asking very carefully how confident we are that the person we might vote for genuinely cares about all the other people in their country and in the world. We barely seem to talk about this incredibly important characteristic. Actually, I think honesty and kindness usually go pretty well hand in hand; if you care about other people, you can afford to be honest with them and transparent in your dealings, because everyone will see from the way you behave that you care. If politicians are operating out of self-interest instead, then transparency and honesty become much more challenging for them because they are bound to be trying to persuade people to do things that may not be in their best interests. Please feel free to add kindness to your own selection criteria.

Finally, in these five honesty criteria, I have not even included competence, although it is clearly also important. I think it is slightly *less* critical than the fundamentals of honesty and kindness. If someone lacks those two qualities, I would rather they were incompetent. On both sides of the Atlantic and around the world, I can think of people who I wish were less competent. What is incredibly dangerous is when someone who is dishonest and self-serving also turns out to be charismatic and smart. These are the people we most need to get better at guarding against.

What Can Politicians Expect in Return?

Respect needs to be mutual. Just as politicians shouldn't abuse the public with dishonesty, so the public needs to get used to treating politicians with a lot more respect – even when they don't keep their side of the deal.

For a start, we stay away from their personal relationships, their families and their homes. They and their families should feel safe. Next, since being a good politician takes a lot of hard work and since we don't expect them to be on the take from those who might tempt them away from acting in the public interest, I'd be in favour of paying MPs and ministers more generously for the basic job that they do. This is consistent with a more equal society. Being a national politician – or at least being a good one – involves sacrifice, and we should reward it.

We should be polite and respectful with them, even when we are furious with them or when we think they have let us down. Corrupt politicians are still sentient. We should allow them to take their time over important decisions, wherever this is possible, and we should allow them to be uncertain in the meantime. We should be careful not to misrepresent them, nor to consume media that disrespect them. We should allow them to present more nuanced and complicated arguments than simple sound bites because the issues we need them to deal with are complex.

When Is It OK for a Politician to Change Their Mind?

Whenever they have good reason to. And we should encourage them to do so as long as there is a good explanation and, where necessary, sufficient humility and evidence of learning.

There are two very good reasons for changing your mind. The first is that the situation, or the evidence upon which the original decision was made, changes significantly to make the arguments point in a different direction. When this happens, it is wrong *not* to change your mind. If the evidence points things in a 180 degrees different direction, then the responsible and trustworthy politician does a U-turn, explaining the reasons why. And responsible media applaud them for it. And a responsible public respects it too. (This is of course different from U-turning on something for illegitimate reasons and concocting a storyline pinned together by convenient facts and/or fictions to justify it.)

The second reason is when you realise that you have been wrong. When this happens, the right thing to do is to change your mind, explain how your thinking has changed, and do so with great humility. Done this way, responsible media and public will also respect you, since we all have to change our mind sometimes as we learn. In fact, in a major survey of UK citizens, fielded by University College London, owning up to mistakes came up as the second most highly valued attribute for a politician.[14] But you can also expect, quite rightly, that questions will be asked about your judgement and the extent to which you have learned from any mistakes made, in order to give confidence in your future judgements.

How Should I Treat a Politician Who Has Abused Me with Dishonesty?

With respect, firmness and clarity.

One of the toughest challenges is to apply the principle of universal respect even to those whose behaviours make our blood boil. The line to take is not 'You're nasty. Get out of politics' but 'The evidence is that you do not have the qualities – namely, honesty and integrity – that we require from our politicians. Therefore, you are unsuitable for the job.'

It's tricky. First of all, we have to find a way of genuinely holding that respect, even for those who make the world a worse place and may never have the capacity to repent. We need somehow to respect those who do not have the self-awareness to understand their negative impact, and those who lack the empathy to care. The principle of respect has to apply to *everyone*: psychopaths need containing but not disrespecting. Respect does not equal weakness.

Two examples come to my mind of the power of this approach. The first is Nelson Mandela's incredible skill in avoiding a bloodbath as apartheid was dismantled in South Africa. Without condoning the actions of the past, he managed to find a way of respecting his adversaries as humans. The second example is the disarming power it gave to Extinction Rebellion in the spring of 2019, when they blocked bridges and roads in central London. They repeatedly emphasised their respect for all; not just the public, but also the police and even those running fossil fuel companies. 'Love and rage'.

It is so tempting to get drawn into the fight. The high ground is so much more powerful.

What Does It Take for a Politician to Regain Trust?

It's tricky but possible.

Firstly, time on its own isn't enough. A person's truthfulness doesn't automatically improve just because we forget and move on. So, we need to make sure we have long memories about those who have deceived us. If someone has been sacked for lying as a journalist,[15] 20 years later the default position has to be not to trust them and to question the judgement of anyone who does. The reason we reject a politician for carelessness with the truth is not in order to mete out a punishment. It is because we have discovered that they lack a vital quality for the job they are in and are therefore unfit for the role. It is essential that our decision stands until we have evidence to justify reversing it.

For example, Jonathan Aitken MP was given 18 months in jail for perjury in relation to hotel bills that were being paid by a Saudi prince. He 'found Jesus' in prison, is now a priest and campaigns for prison reform. Does this mean he can now be assumed to be trustworthy? He had plenty of time to reflect, and his religious conversion may have given him the opportunity, at least, for a major reset of values. I'm not making the case that Christianity or any other religion necessarily makes a person honest. Scandal after scandal has told us that being a priest is not proof of honesty. Peter Oborne, creator of Political-Lies.co.uk, is clear in his book *The Assault on Truth* that 'Proven liars can never be trusted. Someone who lies once will lie again.'[16] I'm prepared to be more open than this to the possibility of change, because I think there is evidence that all of us can change our values and habits by deliberate effort. So, I'm up for cautiously giving Aitken a second chance. But the point I'm making, in using him as an example, is that we need to see evidence of regret and a believable mechanism of change in order to trust again someone who has lied to us.

How Should I Vote If All the Candidates Are Unsuitable?

Vote for the best one that stands a realistic chance. However bad you think they all are, the difference between them matters just as much. Your vote pushes for a higher standard.

It seems there is no limit to the depths of unsuitability that it is possible for our politicians to descend to. However terrible all the options might look, it is essential to remember that they could still get much worse. Think of wars and genocides. Wherever you think we are on the continuum from wonderful to awful, if you pick the best of the bunch you push things in the right direction. It is imperative to vote even if all the candidates are terrible. Otherwise, we deserve it when things get even worse.

Why Is It Self-Defeating to Vote out of Self-Interest?

The change that your vote makes to your own life is so minute that it only makes sense to bother if you vote to improve the lives of everyone else as well.

The chances that your vote alone will change the election of even a single politician are tiny. In fact, in the UK, going back right through the records to 1918, when women were first given the right to vote, it has never happened that even a single MP has won their seat by just one vote. And yet we have to go to quite a lot of trouble to get to the polling station, into the ballot box and back home again. It takes most people at least a quarter of an hour to go through the process, even if you set up a postal vote. And that doesn't include any time you might spend considering the issues and thinking about who to vote for.

Let's do the sums to see if voting is an effective way of serving self-interest. In the UK there have been 29 general elections since 1918. There are currently 650 constituencies, although the numbers have varied a little. There have also been some by-elections. All in all, there have been around 20,000 MP election opportunities since 1918. Even if you had cast a vote in every constituency in every one of those elections over the past century (which of course is impossible), and if you managed to take only 15 minutes per vote, you would still have spent around 5,000 hours voting in total – equivalent to around three years of a nine to five job. And you still would not have changed a single MP. And even if you had, although you might have felt some modest benefit in your personal circumstances, you probably would have had minimal effect on the whole government. So why does it make sense to bother voting?

The answer is that when we vote, if we are smart, we do so in the interests of everyone else as well. Voting makes sense when we do it to improve everyone's lives as well as our own. Only when viewed in this way does it become worth the bother. In the UK, such an attitude makes your one vote 68 million times more worthwhile than if you were only thinking of yourself. And you can make your vote an even better way of spending time if you keep in mind the best interests of the whole global community of eight billion people. That is the only mindset that makes sense at the ballot box.

So, voting only makes sense as an act of kindness. Voting selfishly is a waste of time. If self-interest is your mindset, both you and the rest of the world will be better off if you stay at home and put your feet up instead.

What Democratic and Parliamentary Reforms Would Help?

You don't have to be an expert on democratic and parliamentary procedure to see that few, if any, of the world's so-called democracies are set up to deliver the quality of decision-making that we now need. In the UK, things are completely broken. In Appendix 3, I've included a few pages of suggestions on how things might be improved, pulled together from a range of very smart people who are closer to the workings of politics than I am, along with a few of my own thoughts, benefiting from the fresh eyes of a relative outsider – which I hope is also useful, given how much needs to change.[17] Although these suggestions are specifically focused on the UK, much of their content translates to other stumbling democracies around the world. Here in the main text, I'll just outline a few very quick principles.

Every electoral system needs to be fair and inclusive, and to adequately reflect the informed will of the people. This means doing away with crazy madnesses like first-past-the-post voting, which favours parties that don't have their votes split by rivals with similar ideas, and forces the electorate into depressing tactical voting games that are a long way away from the free expression of their democratic will.

All politicians need to be held to very high standards of integrity, roughly along the lines of the Nolan Principles, and in return they should be treated with respect by each other, by the public and by the media. They should be comfortably paid and expected not to be raising money on the side by means that are either distracting from their day job or compromising their integrity.

Honours and public appointments need to be based on merit and not a reward for loyalty to the party making the appointments, nor simply seen as a means to promote a party-political agenda.

The public needs to be involved and engaged throughout the political process and not just at election time. The decision-making processes need to enable high-quality decisions on complex problems. There needs to be thoughtful, joined-up thinking, honouring the best available evidence and respecting the many stakeholder interests. This will require a huge evolution from the traditional adversarial debate that, for example, is still taught in some of today's schools and is business-as-usual politics in the UK, the US and elsewhere. Citizens' Assemblies have proved to be one powerful mechanism for ensuring public representation and careful, evidence-based deliberation (more on Citizens' Assemblies in Appendix 3).

Citizens' Assemblies and Citizens' Juries

These are proving to be an important and exciting innovation in the democratic process. A group of the public, carefully chosen to represent the whole of society as closely as is practical, is given time, facilitation and access to expert briefings to understand a specific topic. They are given plenty of opportunity to ask questions, to discuss things amongst themselves and to seek further sources of whatever expert advice they request, with a view to producing a set of recommendations. While the UK's Brexit referendum told us what the public thought once it had been subjected to a blend of information, misinformation and every campaigning tactic imaginable to manipulate its opinion, a Citizens' Assembly is designed to reflect how the public *would* think if everyone had had the opportunity to consider the issues very carefully together, free from partisan persuasion and hype, and in the light of the best available factual information; taking time to hear and understand each other's insights and perspectives. In Ireland, they have now been used many times to help unlock thorny issues including abortion, an ageing population, climate change and even changes to the way their democracy functions.

The UK's Climate Assembly in 2020 was commissioned by six Select Committees of Parliament to explore the UK's path to net zero. It involved 108 people and took place over six weekends.[18] The process was carefully facilitated and monitored by independent observers, as well as by the press. It was all designed to bring out the highest-quality collaborative decision-making from a diverse group representing the whole of the UK adult population. I was one of about 30 so-called 'expert witnesses'. It was an impressive event to play a small part in. Walking into the room was like being put in front of the whole of the UK, in all its diversity, condensed down into about one train carriage-full of people. Like all the other witnesses, I had been given a tight brief for a short presentation, after which I was sent

> out while the group discussed what they'd heard and came up with their follow-up questions, some of which I then did my best to answer to the whole group, and some of which I answered in discussion, moving between small tables along with other expert witnesses. Every table had a facilitator. Great care had been taken to set up an environment that encouraged those who might not normally take part in these kinds of conversations, or who might be less confident, to participate as fully as everyone else. The output from the Climate Assembly was incredibly sensible, covering recommendations of all the main components of the UK's journey to its legally adopted target of net zero by 2050. The only problem with the process was that it was largely ignored by government, and there wasn't much to stop that from happening. The missing piece in the puzzle was something to make it incumbent on ministers to either act on the findings or present a robust rationale for not doing so, backed, not by bullshit, but by a process that had demonstrably more rigour than the Assembly itself.

Transformation of the political culture and processes looks to me like the easiest and most powerful lever to pull for all those wanting to see progress at last on the Polycrisis. The changes we need are not possible without it, but political change is also inextricably bound to the need for radical improvements in our media. So, we'll look at that next.

8 GETTING TRUTH INTO THE MEDIA

The explicit and implicit messages from our media sink into us almost no matter how hard we try to defend ourselves against their influence. Both through their content and the adverts they run, they affect what and how we think. They are skilled at making us believe what they want us to.

Much of the media (both traditional and social) does not care about what is true, about your best interests, or about the best interests of people and planet. Unless you actually want to be played, such media need to be starved out. Please do not give money to them, do not trust them, do not consume them (except, with extreme vigilance, to learn about what they are saying, rather than what to believe). But that is not enough. If we want to change media culture, we have to be prepared to expand our impact by challenging friends, colleagues and relatives who support or absorb rubbish media.

Recruiting for a senior executive position in my company recently, we asked the candidates how they got their news; how they chose their sources and what their basis was for working out how much they could trust it. With one or two exceptions, it was shocking to hear some thoughtless responses from otherwise smart people. It is through the media that we gain most of our understanding of current world events, and of our politicians. Even more than that, the news media push their values onto society: through the lines they write, through the suggestions and flavours that emerge between those lines, and in what they choose *not* to cover. If our media pre-suppose that GDP growth is what matters most, then we will adopt that value unless we hear otherwise from a source we take more seriously. If we listen to our national radio every morning and the Polycrisis is treated as a peripheral issue that doesn't really affect what all of us should be doing right now, then it takes a lot of strength to remember – day in, day out – that it is much more important than that. If a newspaper suggests to us that dishonesty is just how the world works, then unless we are unusually independently minded, that – over time – is how we too will come to view society. It doesn't have to be a direct statement that lies don't matter. In fact, those are much easier to spot and challenge than, more dangerously, when that value is pushed obliquely – perhaps by endorsing a candidate who is known to lie, or by treating a misleading statement as if it is just one issue among many on the agenda.

Why would anyone read a paper that has a track record of dishonesty and whose owners push a political agenda for personal gain? While I understand the arguments for staying on X to have a counter-voice, as a source of information it is now deeply dangerous. Elon Musk has, for example, crawled all over the algorithms to push as hard as he can for the election of a convicted felon, everyday liar and self-confessed 'pussy grabber' to be US president. It is frustrating for me every time I see someone buying a paper that is known to be dishonest and driven by profit rather than by the best interests of the reader. But at the same time, there is a price to be paid for calling out the press. I would love all newspapers to give this book glowing reviews, but it feels unlikely, given that I am going to be discouraging everyone from reading the majority of the UK's media. The pressure on politicians is greater still. How can a party leader afford to take the risk of upsetting such powerful elements of the press? Can we blame them for sucking up to the Murdoch empire, which has a convincing claim to have won elections for its preferred party? It is fundamentally corrupt for politicians to do deals with the media, but also tempting, if the price for not doing so is seen as electoral defeat. As things stand, in the UK, the USA and elsewhere, it is extremely difficult to be elected against the will of media that are owned by a small number of billionaires, all with vested interests that are not fully aligned with the public good. The idea, for example, of reducing the gap between rich and poor stands in contrast to the financial interests of the owners.

In the UK, when the Murdoch empire was in trouble over the phone hacking scandal, Michael Gove helped to bail them out under the guise of defending 'press freedoms', and Murdoch reciprocated in part by supporting Boris Johnson's Conservative government.[1]

Staying in the UK, if the BBC isn't passionate about honesty, then it is much harder for our society to have high standards. In this chapter, we are going to look at how to assess the trustworthiness of the media and what to do to improve it. I will even make a few specific suggestions about some sources that I think score highly against these criteria and some that don't. You might like or dislike these outlets for other reasons, but all I am assessing is their honesty. My own view, however, isn't really the point. What I really want is a world in which everyone is scrutinising their own criteria for picking their media, so that the basis on which they are selecting their sources is at least as strong as what I'm putting forward here – or better.

To repeat, much of our media is owned by billionaires and exists primarily to make money and/or serve the owners' personal agendas, regardless of the public interest.[2] Rupert Murdoch, owner of the *Times*, the *Sun* and Fox News, even believes that climate change is not much to worry about.[3] These organisations do what it takes to gain readers and viewers and to push their owners' agendas. If you read them, you are being manipulated for the purposes of the owner.[4] Even if the facts are correct, your understanding of the world is being steered by the stories they choose to put in front of you and by the tone in which they are written. If you pay them money or view their adverts, you are supporting them. If you read their content, you are absorbing the world view that they want you to have, and you risk being taken for a ride.

Here are just a few of the messages that soak into us through the background assumptions of most of our media, but that don't bear scrutiny:
- The Polycrisis is peripheral to our daily lives.
- Conventional attitudes to growth, including the central importance of GDP and the necessity of rising energy demand, are not to be questioned.
- The primary purpose of business is profit.
- It is natural for a politician to seek more power in all circumstances.
- Our role in society is to consume and thereby to make society richer by growth in the economy (rather than our role being to contribute to society as citizens).
- The lives of people in our own country, or our own 'tribe', are more important than the lives of people elsewhere.

How Can I Tell What Media to Trust?

Here are some simple criteria that you might use or adapt:
(1) **Ownership and governance.** Who owns them? Does the owner practise the values you would like to see propagating in the world? Do the owners have a track record of integrity? It is unrealistic to hope for a 'Chinese wall' between the owners and the editors – in the end, journalists' and editors' careers depend on following the agenda of the owner.[5]
(2) **Funding.** What is their revenue stream, and does this leave them editorially compromised? GB News, for

example, relies on funding from billionaire hedge-fund manager Sir Paul Marshall and the investment firm Legatum. The Conservative Party also helps it by being its largest buyer of advertising.[6] It is difficult for a media outlet to run a story that is critical of a major advertiser or any other revenue stream. The BBC, meanwhile, gets most of its money from the government, which has continually been exerting leverage by threatening to withdraw the licence fee. The *Byline Times*, on the other hand, is funded almost entirely by readership donations. One way of assessing a media outlet is by looking at the companies that they are prepared to advertise.

(3) **Track record.** What is their track record for factual accuracy? Have they ever deliberately deceived the public or knowingly been careless with the truth? If so, what evidence is there of change since then? If they make a factual error, do they issue a correction? What is their track record of taking on the stories that are most significant for the world? Have your social media prioritised click-bait over reality? Have they gone to enough effort to delist inaccurate and fake content? Or have they been a vehicle for misinformation and/or election-fixing?

(4) **Values.** Do they promote values that are good for the world, and not just for a wealthy minority? Do your social media police illegal content that breaches these values?

(5) **Transparency.** Are they transparent with their information sources and are they high quality? Are the social media algorithms by which content is pushed your way fully transparent?

(6) **Third parties.** What do other sources that you trust say about them? This of course only works if you have applied a good enough process for working out whether to trust those sources.

What If I Enjoy Media That Fail These Tests?
Don't read them or watch them.

It is not worth it. It doesn't matter how entertaining it is, or how great the TV guide is, you will be under its influence in everything you read or watch: the comedy shows, the recipes, the film reviews, the travel section, the crossword clues, the adverts: *everything*. If the media owner doesn't want you to be on top of the Polycrisis, one way or another, everything in the content will be pushing you (mainly not explicitly and mainly without you noticing) to carry on as if it isn't happening. The media we absorb and support are a serious matter. If they don't have the values that we need to cultivate, it is important not to be under their influence and to starve them out.

It might seem innocent to watch an entertainment show on Fox, for example, but in doing so you are propping up the Murdoch empire's advertising revenue. Even more importantly, it is a highly unusual person who can do this without being influenced by its underlying values, narratives and social norms.

How About Podcasts and Social Media?

Overall, does society have a clearer view of the truth because of social media? Probably not. On average, we are misled at least as much as we are informed.

Social media give us the potential for greater choice of our information sources, but if we are not careful they lock us into a self-reinforcing bubble, especially if we allow our social media to make recommendations based on what we are already looking at. How to avoid this? Pick very carefully who you read, listen to and watch, based on a thoughtful assessment of their trustworthiness. Keep that under constant review. Try to expose yourself to sources that meet your criteria even if their views are different from your own. Ignore adverts and anything the media channel recommends for you, unless you can verify it is from a trustworthy source. Pick who you follow using information from outside your social media bubble.

Podcasts can be wonderful sources of information, but also dangerous echo chambers. If they pepper their content with adverts and paid endorsements, ask yourself carefully what their choice of these tells you about their integrity. Would they be brave enough to run a story that is critical of a sponsor or major advertiser?

Malign influences invest in both harvesting data from social media and in creating targeted fake content to corrupt democracies. Most people like you and me think they are too smart to be influenced, but these corrupters don't spend their money in this way in order to waste it: they do it because they know it works. Do not underestimate how effective and influential these tactics are, and how easy it is, for all of us, to be duped. Misinformation on X was instrumental in stoking the UK's 2024 race riots. Elon Musk rapidly corroded what used to be the much more respectable Twitter by cutting its links with the Trust and Safety Council and even added his own false and inflammatory post that civil war in the UK was inevitable. He used it to push hard for Trump's victory in 2024.[7] (I'm in transition from X to LinkedIn, with a splash of BlueSky.)

Facebook and Instagram (owned by Meta) derive the vast majority of their income from advertising,[8] which seeks to persuade you to buy things regardless of your best interests or those of the planet. Facebook has also been widely accused of failing to take down poisonous content before it goes viral.[9] In 2014, Facebook data was improperly used by Cambridge Analytica to build up voter profiles in order to influence US and UK elections.[10]

A Tour of Selected Traditional UK Media

This is not a definitive list of what to read and what to ditch, but just a few key examples and personal thoughts, based on the criteria above. At the risk of bad reviews for this book, I'm going to name some popular publications that I hope you won't touch with a barge pole. By that I don't *just* mean don't give them money, I also mean don't support their advertising revenue by consuming the free versions; don't allow them to influence you through their copy. If they can't be trusted to be honest with you, you know they don't have your interests at heart.

I'll begin with some bad media, but then list a couple of good ones, so we have something to look forward to.

The ***Times*** and ***Sunday Times***, the ***Sun*** and **Fox News**. All of these are owned or controlled by the Murdoch empire which has been successfully pushing a hard-right agenda on both sides of the Atlantic. Rupert Murdoch, alongside direct editorial interference, has personally assisted Trump, Thatcher, Bush, Blair and Johnson into government.[11] In the UK, his papers enthusiastically pushed for the Iraq War and supported it long after it proved disastrous. They had a pivotal influence on the Brexit referendum. In the US, Fox News loudly called George W. Bush's election victory without the evidence (which is still unclear), and even more seriously promoted the bogus claim that Trump won in 2020.[12] The *Times* still has a regular column for Matt Ridley, who has denied and later downplayed the climate emergency for many years.[13] You might ask whether a paper that has been so careless-at-best with the science as to give him column inches can be trusted on anything that matters.

David Yelland, former *Sun* editor, admitted, 'All Murdoch editors, what they do is this: they go on a journey where they end up agreeing with everything Rupert says but you don't admit to yourself that you're being influenced. Most Murdoch editors wake up in the morning, switch on the radio, hear that something has happened and think: what would Rupert think about this? It's like a mantra inside your head, it's like a prism. You look at the world through Rupert's eyes.' And Rupert Murdoch himself confirmed this at the Leveson Enquiry into phone hacking by saying, 'If you want to judge my thinking, look at the *Sun*.'[14]

> ### Beyond Phone Hacking?
>
> ***Was there also widespread political and commercial espionage at the Murdoch empire?***
>
> Nick Davies is the journalist who in 2010 unearthed the phone hacking scandal in which Rupert Murdoch's News International was found guilty of widespread and deeply illegal tapping of voice messages. The public was horrified, not least because one of the victims of this spying was a murdered schoolgirl. The scandal resulted, in 2011, in the former editor of the *News of the World*, Andy Coulson, going to prison and the collapse of that paper. Since then, many more of News International's hacking victims have sued for damages, and News International has settled out of court for an estimated total of $1.2 billion.[15] The vast majority of victims don't have the enormous financial resources needed to press *criminal* charges against the mighty and wealthy Murdoch empire (and remember we've seen how loudly money can talk in our legal system). However, these civil cases have led to a wealth of new information being made publicly available. Nick Davies has now used this new information to piece together evidence that looks to me to be very highly suggestive (I'm choosing my words carefully) of further criminal activities in the Murdoch empire; not just of spying but of corporate and political espionage – deliberately targeting through illegal means MPs, including cabinet ministers, either to

discredit them through embarrassing revelations about their private lives or to obtain information about decisions relating to News International's acquisition of BSkyB. Here are some excerpts:

- *News of the World* carefully deleted 31 million emails and destroyed nine boxes of physical evidence and a large number of computers. This took place despite police requests to preserve such evidence. It happened at the explicit and repeated insistence of Rebekah Brooks, now CEO of Murdoch's UK News.
- News International set out to 'Get Chris Huhne' when he threatened to expose Coulson, who had become David Cameron's head of media and an important News International plant into the very centre of government.
- Thousands of highly suspicious very short calls were made from News International offices to MPs that they were seeking to discredit. These calls are not remembered by the MPs themselves, were too short to be legitimate conversations, and came from the News International offices rather than any coming from journalists' own phones, which Nick Davies suggests would be more normal. All this is consistent, Davies contends, with attempts to hack phone messages.

News International denies Nick Davies's allegations, and claims, for example, that there is no proof that the thousands of suspicious calls could not have an (unspecified) innocent explanation. We have to work out who and what to believe. The five criteria that I have laid out for working out whether to trust a politician can also be used to assess journalists and news empires. I will let you make up your own mind. Nick Davies lays out his case in a 15-page article in *Prospect* magazine, which is edited by the highly respected Alan Rusbridger, who presided over an era of top-quality investigative journalism when he was editor-in-chief at the *Guardian*.[16]

The **Daily Mail**, the **Mail on Sunday**, the **Metro**, and now sadly **New Scientist** are owned by billionaire Lord Rothermere (Jonathan Harmsworth), a prominent supporter of the Conservative Party who is 'non-dom' for tax purposes. To give just one example of the *Daily Mail*'s poisoning of democracy, it ran the headline 'Enemies of the people' against the Supreme Court judges who ruled that Boris Johnson's attempt to prorogue Parliament had been illegal.[17] In other words, it was supportive of Johnson's attempt to undermine the rule of law. As recently as 2023, the *Daily Mail* ran an editorial entitled 'Climate hysteria', dismissing the IPCC's latest report.[18] This tells you all you need to know about a paper's judgement, not just on climate but on anything at all that you might care about. It doesn't even matter whether or not you or I might be able to find other examples of times when we think the *Daily Mail* may have undermined truth and democracy, because the wider point is that just one good example is all you need to know when deciding whether it is OK to buy this paper or even to stay quiet if you know anyone else who does.

It would be wrong to single out the *Daily Mail*.

The simple knowledge that the **Telegraph** employed Boris Johnson as a prominent journalist, despite his clear-cut track record of dishonesty in both journalism and politics,[19] is, in my view, enough on its own to blow its credibility out of the water. Ownership gives a further reason not to trust either that paper or its sister publications. The **Telegraph**, **Sunday Telegraph** and the **Spectator** are owned by the UKIP- (UK Independence Party) and Conservative Party-supporting billionaire Frederick Barclay, who minimises his tax contributions by living on the island of Sark, and against whom – according to an investigation by the *Economist* – there are 'strong grounds to believe' his business empire has been built using both fraud and tax evasion.[20] As if more reasons not to read this paper were needed (which they are not), here is just one example of its red-card-worthy content. The Independent Press Standards Organisation forced the *Telegraph* to correct multiple false claims in an article it published in 2019 in praise of Trump's inaction on climate change.[21] The headline (which is *still* on their website) runs: '*Donald Trump has the courage and wit to look at "green" hysteria and say: no deal: The iconoclastic president has broken the spell of climate change mania.*'[22] The author is Charles Moore, former *Telegraph* editor and a trustee of the appallingly dishonest Global Warming Policy Foundation.[23] (At the time of writing, ownership of the *Telegraph* and *Spectator* is up in the air after the government blocked the attempted sale to a Gulf-states-backed consortium.[24] To give credit where it's due, the UK is introducing legislation to ban foreign governments from owning UK newspapers.)

The **Daily Express** and **Daily Star** are controlled, according to the *Guardian*, by the Conservative donor, less than scrupulous lobbyist and former pornographer Richard Desmond.[25]

If we want humanity to stand a chance of thriving in the Anthropocene, all readership of the* Daily Mail, *the* Telegraph, *the* Times, *the* Sunday Times, *the* Sun, *the* Express *and the* Daily Star, *for example, needs to be challenged.

Now for some examples of media with much better track records of truth – note, sadly, there aren't quite as many . . .

The **Guardian** is a not-for-profit B Corp, overseen by the Scott Trust, which has been carefully set up to ensure editorial independence. It is funded through endowments and reader contributions. There is some paid content which is very clearly labelled as such. Digital and print subscribers receive their content advert-free. It has a very strong history of investigative journalism and usually cites its sources, which generally turn out to be robust. Overall, in my view, they have a good track record of care with the truth. Errors are visibly corrected, and there is a clear process by which readers can highlight possible errors for investigation. Some people criticise the *Guardian* for being 'left-wing'. I wonder how much of that is because when you look at the evidence without the influence of a billionaire owner, it takes you to a more egalitarian position? It makes it clear when you are reading factual journalism, and when you are reading opinion – not all of which I agree with, but I don't have to. Interestingly, when the *Guardian* crops up on fact-checking platforms it is much more likely to be fake stories about what the paper is saying, rather than errors made at the *Guardian*. The *Guardian* has been B Corp certified, and although I'm often sceptical of these kinds of corporate badges, B Corp is worth something in my view. Although the *Guardian* bans fossil fuel advertising, it still allows harmful adverts such as by airlines. Finally, and at the risk of taking me into a thought-bubble, the *Guardian* is also rated by many smart people whose integrity and judgement I trust.

Financial Times: A for-profit publication, owned by the Japanese company Nikkei, which, in turn, is owned by its employees. Both the FT and Nikkei stake their entire reputation on the trustworthiness of their content, for which they are both highly regarded around the world. The FT Editorial Code requires all contributors to report any errors they subsequently find in their work, to respond to queries that arise, to take great care with their sourcing and attribution, to declare interests and more besides.[26] To my mind, it is trapped in outdated traditional economic thinking that we have to learn to get beyond. But with that caveat, and provided, as you read, you are able to pull out and correct for the presuppositions about how the world has to operate, it is a reliable source.

Byline Times: A monthly paper, also available online, funded almost entirely by its readers and founded to be free and independent, and a counter-balance to the billionaire-dominated British press. It has a very credible set of journalists, working for it because they were sick of the lack of truth in mainstream media. Look them up and make up your own mind. It often tells me things I didn't know, but I can't think of an example of it having been factually wrong. I would love it if it cited its sources more.

There are more examples of media that contribute to a thoughtful understanding of the world in Appendix 4.

How Is the BBC Getting On?

The once world-renowned BBC is still publicly owned and publicly funded. However, it has been under sustained attack. Almost 20 years ago, Dominic Cummings, former senior adviser to Prime Minister Boris Johnson, while director of a right-wing think tank (the now-defunct *New Frontiers Foundation*), called for an end to the BBC 'in its current form', and replacement with a model that was free from impartiality rules. He proposed the deliberate undermining of its credibility and its funding model, and that government ministers should avoid appearing on its news programmes.[27] Much of this advice was progressively implemented by the Conservative government between 2010 and 2024. He also recommended setting up a right-wing rival to 'shift the centre of gravity' of news to the right, and sure enough, the ultra-right-wing GB News has emerged. It follows the Fox model and has so far avoided significant punishment by Ofcom,[28] despite repeatedly breaching broadcasting impartiality standards, for example by Conservative MPs hosting programmes in which the guests are also right-wing politicians.[29] BBC funding through the licence fee has been severely cut by Conservative governments, and the threat of its removal altogether has served to further cow the BBC.[30] Meanwhile, senior positions have been filled with Conservative Party members and donors: Director General

Tim Davie is a former Conservative candidate;[31] Robbie Gibb, a former director of communications for Conservative Prime Minister Theresa May, is now on the BBC board with a specific role in safeguarding impartiality(!). Richard Sharp was appointed Chairman of the BBC despite being a close confidant of Boris Jonson and a committed Conservative. In that role he exerted considerable editorial influence, including steering the appointment of senior editors (he was ousted when it was revealed he had arranged for Johnson to be lent £800,000). It all stinks. Alan Rusbridger, former editor of the *Guardian*, and now editor of *Prospect* magazine, has written in detail about what he concludes to be concerted and scurrilous attempts to engineer government poodles into key BBC roles, despite their clear lack of credentials for those roles.[32]

The impact of all this on the BBC's content is plain to see. Misleading statements from Conservative government ministers have been allowed to pass almost or entirely unchallenged. False statements in the Brexit campaign were allowed to stand, resulting in a seriously misinformed public on referendum day. Arguments are presented as two equal sides of the coin, even when one side is provably bogus. As recently as 2017, climate change denialist Nigel Lawson was brought onto BBC Radio 4's Today programme to 'balance' the scientific consensus.[33] Climate change is still treated as a peripheral issue. The once-proud BBC failed in its duty, year after year, to call out a serial liar running the country, even though Channel 4 managed to do so.[34] The BBC gives undue airtime to bogus think tanks, whose funding streams are opaque and whose honesty track record tells us they should never have airtime. Claims by the 'Common Sense Group' of Conservative MPs that the BBC exerts left-wing influence are in direct contradiction to more impartial research looking into this very issue.[35] The routine dishonesty that I have touched on earlier in this book, and which poured out of Rishi Sunak's government in its final months, went largely unchallenged.

Where does this leave us? The BBC is not yet completely useless, but now needs taking with a large pinch of salt. I trust Channel 4 somewhat more – which is probably why the Conservative government tried to get it privatised in 2023. Please don't switch to any news outlet owned by a billionaire with a political or personal agenda.

It's easy to feel depressed by such media problems, but the good news is that we don't have to stand for it.

Bogus Think Tanks

My criteria for assessing the trustworthiness of politicians, and the Nolan Principles of standards in public life, are also applicable to think tanks. Using these criteria, I would describe many think tanks as bogus. What I mean is that if they lack transparency of funding, use dodgy information sources and/or have a track record of misinformation, they are bogus.[36]

The words 'think tank' imply that careful, objective thought goes into the advisory reports they produce, but this implication can be as misleading as the titles they give themselves – which seem to me often to be designed purposely to pull the wool over people's eyes. For example, the 'Global Warming Policy Foundation' actually denies a lot of the science around global heating. 'The Tax Payers' Alliance' argues against paying more taxes and promotes itself as a grassroots movement standing up for the UK tax payer against wasteful government spending. It describes itself as 'non-partisan'; a little digging, however, reveals a clear libertarian agenda and corporate interests at play.[37]

They are both part of a large handful of highly organised and interconnected think tanks, with opaque but substantial funding, some of which have been pushing out climate denial and confusion with great effect. The Global

Warming Policy Foundation, founded by the infamous climate-denying Nigel Lawson, is perhaps top of the list for its destructive influence. As of 2022, GWPF received funding through the US Charity 'American Friends of GWPF' whose donors include the libertarian billionaire Charles Koch, who is also highly influential in pushing US politics to the right, also through think tanks.[38] Others include **Civitas**,[39] which denies the science on climate change and produces 'educational' material,[40] the aforementioned pro-free-market Tax Payers' Alliance[41] and the **New Culture Forum**.[42] All these are co-housed at 55 Tufton Street, just down the road from the Houses of Parliament. Next door at number 57 is the **Centre for Policy Studies**.[43] The **Institute of Economic Affairs**,[44] a two-minute walk away, receives annual funding from BP, as well as gambling and tobacco interests, although most of its funding is chillingly anonymous through The Donors Trust – described by Mother Jones, America's longest established investigative news organisation, as a 'dark money ATM'.[45]

The **Adam Smith Institute**[46] has also published numerous reports confusing the climate science. The **Institute for Free Trade**, launched by Boris Johnson, which has now moved from 57 Tufton Street, is focused on deregulation, but also downplays climate change with the bogus claim that we can just adapt to it with a 'fractional sum of money'.[47] The IFT teamed up with the **Cato Institute** (funded by the Koch brothers, plus oil and tobacco industries) to push for deregulation of trade between the UK and USA, including opening up the NHS to foreign competition.[48]

One way or another, many millions of pounds find their way from oil companies and over-rich libertarian donors into these highly influential bullshit-purveying

> think tanks. I can't detail it all here, but the endnotes for this chapter contain links to articles by some of the credible and financially transparent organisations that have managed to unpick some of how this system works: OpenDemocracy, DeSmog, the Grantham Institute at the London School of Economics, Mother Jones and the *Guardian*.[49]

What Can I Do to Improve the Media?

Four simple suggestions which apply equally to both traditional and social media:

(1) Choose your sources extremely carefully, applying the criteria listed above, or your own better criteria. Try to have varied sources. You don't have to follow my specific recommendations and blacklists, as long as your own selection process is as robust as mine or better.

(2) If you can afford to, pay for any good media that you absorb, and consider donating extra. Good media cost money. If we don't want our media to be doing whatever it takes to gain viewers, listeners and readers, and if we don't want them to be powered by adverts whose purpose is to persuade us to think and want things, regardless of whether that is in the best interests of people and planet, then we need to pay for them – and properly. In the US, for example, National Public Radio (NPR) is funded entirely by public donations, enabling it to be advert-free and more impartial at least than the main alternatives. If you give money to any good causes, supporting quality not-for-profit media could be one of the best, or even the best, donations you could make. Similarly, avoid giving money to bad media, and do not support them even by going online

and thereby being exposed to the advertisements they are paid to carry. Bad media must be starved out.

(3) Multiply your impact by challenging anyone you know who absorbs corrupt, truthless or democracy-degrading media. This doesn't have to be about losing friends and falling out with your relatives and colleagues. You can be friendly but clear. We have to have a culture in which reading truthless media is seen as being for suckers who don't mind being abused, manipulated and taken for a ride. I know it can be awkward and boring to have to do this, but we need to eradicate bad media. We can't rely on politicians to do so, because bad media and politicians scratch each other's backs.

(4) Consider actively contributing to the good media through your work, either as a journalist or in some other capacity.

The same principles that apply to the media also apply to companies. Remember, if a media source or business isn't honest with you, it is abuse.

9 GETTING TRUTH INTO BUSINESS

In *The Emperor of All Maladies: A Biography of Cancer*, Siddhartha Mukherjee tells the story of how the tobacco industry played its cards when scientists finally spotted the link between smoking and lung cancer, thus transforming their industry from benign providers of leisure products to mass killers.[1] It's an especially chilling read because the process of defending the industry, by first denying and then deliberately obscuring the evidence, then resisting the regulations that could solve the problem, is uncannily parallel to the actions of the fossil fuel industry over the past few decades in response to the science on climate change. And sure enough, in *Merchants of Doubt*, Naomi Oreskes and Erik Conway unpick in detail, and with widely acclaimed rigour, how a very small but effective group of dishonest scientists stifled global action on climate change. Driven by their belief in free markets and their fear of communism, and handsomely funded by conservative think tanks and corporations with vested interests, they set out to 'discredit the science, disseminate false information, spread confusion, and promote doubt'.[2] The same group worked to perpetuate not just smoking, but also acid rain, the use of DDT and the use of ozone-depleting aerosols, and, most devastatingly of all, laid down the template for the fossil fuel industry's ongoing tactics. That template for defending an industry against inconvenient truths is so effective that even after all these years of scientific clarity, fossil fuel extraction continues to rise. And it is so generic that it can be adopted by almost any industry that seeks to defend its profits against evidence that it harms people or planet. Most recently on my radar, elements of the corporate meat industry have rolled up their sleeves to confuse the debate on sustainable food production.[3] Their work is made easier by the complexity of the science in this area.

Corporate dishonesty, of course, takes many forms. It can have very big money behind it, because the financial rewards can be enormous, and the job of uncovering it can be overwhelming. Those brave enough to take it on are put up against extremely well-paid lawyers who are out to ruin them. We've already looked at how the legal system can be a powerful bullying tool for those who can afford it.

As I write this, the case of widespread, calculated, long-standing and extremely serious dishonesty at the UK Post Office and Fujitsu is all over the UK news and public consciousness. Everyone is asking why this story did not erupt years ago. Hundreds of sub-postmasters have been wrongly accused of false accounting and theft, resulting in them losing their jobs, being humiliated in their communities and being made to pay huge sums of money that they didn't owe. Some have been sent to prison, and several took their own lives.[4] The Post Office and Fujitsu look to have been calculating and systematic at covering up the many embarrassing faults in the Horizon accounting system,[5] which the Post Office had bought at huge expense from Fujitsu. Top executives involved in the cover-up did so in the knowledge that they would be walking directly all over the lives of hundreds of people, and indirectly over the lives of thousands. The evidence of this huge miscarriage of justice has been out there for years: the result of painstaking campaigning from some of the victims, supported especially by some relentless journalism, not least by *Computer Weekly* and *Private Eye*. The BBC got some traction with a *Panorama* documentary in 2015,[6] but not enough to provoke a serious attempt to right the wrongs and hold dishonesty to account.

9 Getting Truth into Business

For years, only a handful of unsafe convictions were overturned, and the chief executive of the Post Office even retained her CBE.[7] Until, that is, an ITV drama, scrupulously careful with the facts but also emotionally engaging enough to bring tears to my eyes, succeeded in making the issues unavoidable for politicians.[8] It is a wonderful example of what the media at their best can do for public awareness and truth. But it has been desperately difficult to get to this point. The issues would never have fully surfaced at all without one or two critical witnesses coming forward among the many who could have done so. Both organisations contained plenty of people who knew at least some of what was going on. Similarly, the secret and illegal parties at 10 Downing Street during the COVID-19 pandemic relied on cleaners, security guards, civil servants and politicians all keeping quiet, when every single one of them could have come forward. The risks, of course, are highest for anyone living hand-to-mouth who cannot afford to lose their job.

How many other cases of serious illegality in the workplace are well known by many staff, but still out there in the long grass? How can a culture of truth be created when it is so hard to expose the dishonesty? Just as when a politician is found to have deceived us, the answer is to keep digging until all the rot has been exposed: we need to see all those who lied and all those who stayed quiet when they should have spoken out. Beyond that, we need to get to the root of the cultural problems which enabled those behaviours. It is tempting to treat the Post Office–Fujitsu scandal as a one-off, but this story and the difficulty in uncovering it are highly suggestive of many more similar stories lying low throughout the corporate world.

At the time of writing, in 2024, countries across the world are engaged in armed conflict – making huge profits for a few in the arms industry while utterly decimating the lives, communities and hopes of millions of civilians. For example, increased military spending, prompted by Russia's war on Ukraine and the Israel–Gaza conflict, helped the British weapons manufacturer BAE Systems to record profits last year, with further growth expected in the year ahead.[9] The FTSE 100 company made underlying profits, before interest and tax, of £2.7 billion on record sales of £25.3 billion in 2023. BAE Systems is one of the world's largest arms manufacturers. When my own company politely declined to work with BAE Systems on the grounds that we found the evidence for widespread corruption compelling,[10] I found myself having to explain the decision to senior management at Lancaster University, in which we have our offices and where I also have an academic role. The university gets significant funding from that company, and BAE's senior management had been on the phone to the Vice-Chancellor about the email in which I declined to work with them.[11] The line of least resistance is always to just pretend these things aren't an issue.

Since the chances of companies getting caught in cases of corruption do not seem that high, the cost of being caught needs to be many times higher than the perceived benefit of getting away with it. (Remember that bit in Chapter 6 where a *climate of truth* was expressed in mathematical terms?) In the Post Office–Fujitsu scandal, so many people played a part in dodgy practice that the whole culture must have been deeply flawed. It is not enough to simply get justice for the victims and hold a few people accountable. Just as with the climate emergency, if we want things to change, we need to get to the bottom of the problem. We need to understand the reasons behind the reasons that this was able to happen in the first place and how it remained unresolved for so long after the facts had surfaced. Until then, we can be fairly confident of similarly awful episodes occurring again and again.

If I try to list every case of corporate dishonesty, this book will turn into an encyclopaedia. Why is it so prevalent, why does it matter so much, and what can be done about it?

It matters so much because the fragile world we live in can't afford the damage to people and planet that it causes. Just as there may have been dishonesty in politics throughout history but now it has to end, the same is true in the business world. We can't afford the luxury of allowing a bit of routine corporate deceit if we are going to stand a hope of surviving the Anthropocene.

Businesses need to exist, directly or indirectly, for the benefit of people and planet.

One of the many problems with fixating on economic growth is that so much economic activity fails to meet this criterion. The investment world has started talking a lot about 'ESG' (Environmental and Social Governance) and a little bit about 'Impact' – a term intended to mean not just being responsible but going beyond that to actually be beneficial. ESG looks like a positive step at first, and this probably explains why it has come under attack from US right-wing interests – resulting in over 160 bills against it in 37 states in the first half of 2023.[12] However, the superficial way in which ESG is often assessed can be so hopeless as to be worth nothing at all, or even be part of the rising sophistication of greenwash. At Small World Consulting (my consultancy company), we've started advising companies and asset managers on the extent to which their own businesses, or the ones they invest in, help to enable the systemic change that we need.

To answer the question as to whether a business is good for people or planet requires a deep and careful exploration of the goods and services it provides, the business model, and its environmental and social impacts. At Small World Consulting, we've found that a meaningful assessment cannot be made without intelligent people taking time to look into the issues on a company-by-company basis. We have developed a holistic framework that combines nuanced interpretation of quantitative data with qualitative analysis. It has been proving useful for both companies and investors. One of the killer questions that we ask is 'If the world economy started transitioning towards sustainability, would this be a net threat or opportunity for this business?' If the answer is that it would be a net threat, then the urgent priority – far more so than climate targets, for example – is to change the business model until the global shift to sustainability represents an opportunity. Until this is done, everything the company does in the name of sustainability will be hopelessly conflicted.

There are some challenging realities here and, to paraphrase Upton Sinclair:

It is often difficult to get someone to believe something when their salary depends on them not believing it.[13]

With this in mind, we'll now take a look at a few key industries, just as examples, to see what it looks like within each for them to be fit for the Anthropocene. The wider point that I hope these examples will illustrate is that the challenges we face today have deep implications for everything that every business, in every industry, thinks, says, does, makes and sells.

Truth in Marketing and Advertising

There is a role for responsible marketing and advertising. MarketingKind is a membership community of marketers and brands who want to use their professional capabilities to create a better world (for example by working on climate change, social isolation and poverty).[14] There is nothing wrong with letting people know about products and messages that are good for people and planet. However, there is no justifiable place for a business or a job that is all about persuading people to think or want things regardless of whether it is in their best interests. If you work for a business like this, it probably has an internal narrative that excuses its practices. You will need to be independently minded to see through this, and then you'll need courage to challenge it. To make things harder, your family might depend on your income. Your loved ones may not see the problem and may be similarly conflicted in their own careers. It is tricky. In the short term, your life is a lot simpler if you don't ask deep questions, but you are reading this book, so you are already doing so.

The UK's Advertising Standards Agency works hard to make sure bare lies are minimised, but that isn't enough to create a responsible, honest industry. António Guterres, the UN Secretary-General, has helpfully called for a banning of adverts by fossil fuel companies, calling them the 'godfathers of climate chaos'.[15] But if you work in advertising you shouldn't need to wait for legislation to tell you not to promote BP, Shell or Exxon. And your responsibilities go much further than fossil fuel. Creating the implicit suggestion that owning a newer, faster and flashier car will make a person more attractive to a potential partner, and therefore happier, is deeply dishonest (you're unlikely to be happy if your partner is only with you for your car). Encouraging people to think only about the effect that a product may have on themselves is also dishonest, in that it is disabling a person from having a full and rounded view of a product and all of its implications. In any advert, a person enjoying driving a car needs to be accompanied by realistic imagery of the extraction of materials and manufacture of the machine, along with all the environmental and social consequences. If women were raped to clear the land to enable mining for some of the minerals used in manufacture, then it is a skewed perspective if this isn't somehow reflected in the portrayal of what it is to buy that car. (And, yes, Amnesty International does claim that this has been going on in the Congo to extract cobalt and copper for batteries.[16]) Similarly, if carbon and other pollutants pour out of the exhaust pipe, then this is part of what it means to buy that car. Does this mean no marketing of cars? No, but a responsible message would include an acknowledgement that all cars have an environmental impact.

So, if you are in the marketing industry, here are your responsibilities:

- Do not encourage people to want products that will not enhance their lives (long term as well as short term).
- Do not encourage people to want unhealthy amounts of things.
- Encourage people to have a full and realistic understanding of what lies behind anything they might buy.[17]
- Try to direct your energy to messages that the public need to become more aware of.
- Challenge things that are not right in your business and your industry.

Truth in Consultancy

This is close to my heart as it's the industry I work in when I'm not writing or researching. The easiest way to make a living in consultancy is to tell your client more or less exactly what they want to hear. Give them the evidence base they need to tell a comfortable story. Often this is cynically done. Often it is more a case of the consultants just not being sufficiently motivated to ask deeper questions.

Four huge firms dominate the global market: PricewaterhouseCoopers (PwC), Klynveld Peat Marwick Goerdeler (KPMG), Ernst & Young (EY) and Deloitte, with nearly 1.5 million employees and US $190 billion in annual revenue.[18] It seems highly unlikely that people and planet are at the heart of their operations. In Australia, PwC is facing criminal investigations after allegedly leaking confidential tax information from the government to its clients.[19] Alongside this, it has been accused by the International Consortium of Investigative Journalists of decades of international scandals relating to cross-border movements of wealth linked to autocratic regimes, oligarchs and tax-avoiding mega-corporations.[20] KPMG was fined a record £21 million in 2023 for its 'very bad' (those words came from the UK boss of KPMG) work on the failed government contractor Carillion.[21] In Germany, in 2023, EY was handed a 500,000 Euro fine and banned from taking on audits for companies of public interest for two years.[22] And also in 2023, Deloitte admitted misuse of government information and conflicts of interest.[23]

These examples are just what they got caught doing in one year, and *if* we assume that each of these transgressions was made following a thoughtful calculation that the chances of being caught would be slim, then these examples could be suggestive, at the very least, of the possibility that other misdemeanours have taken place that have not yet surfaced and maybe never will.[24] All this therefore sounds serious alarm bells and should leave us asking about the extent to which we can trust any of the outputs of these companies. If somebody applies for a job at my company with a background in one of the big consultancies, I'm usually more concerned about what they will have to unlearn, and the values they may have chosen and absorbed, than I am impressed by the experience they may have gained.

Truth in the Fossil Fuel Industry

We owe this industry such gratitude for the critical part it has played in transforming our existence over the past two centuries. And yet the role it has played over the past 50 years has been despicable and devastating. It is tantalising to reflect that if the oil and gas giants had played an honest hand over the course of even the past 30 of those years, we would right now be taking the last of the carbon out of the global economy. Humanity might have taken climate change almost in its stride, temperature rise to date would be well under 1 degree and we'd be on course to top out comfortably below 1.5. The term 'climate crisis' simply wouldn't exist. This industry and the people who have run it have a great deal on their consciences.

But given how things have played out so far, we cannot now turn off all our oil and gas capacity overnight. So, for now, we do still need people to work in this industry, while also pushing hard for its rapid decline. Every employee needs to be clear that increasing renewables is not a substitute for cutting extraction and that only a small residue of fossil fuel extraction can be legitimately 'offset'. If you work for a company that doesn't acknowledge this, then the damage caused by its greenwashing may end up surpassing collusion in any historical genocide that you can think of. And if you don't scream about it, you too are part of that malignant bullshit machine. If there is such a thing as evil, this is it. This isn't me trying to give you a hard time, it's just how it is. If your company produces a promotional video showing your chief executive talking about renewables with a wind turbine reflected in his eye, while failing to mention his plans to expand extraction,[25] and you stay quiet, then you will go to your deathbed knowing that your contribution to history was to have colluded in humanity's greatest ever atrocity.

To help change this industry from within will take incredible strength. The whole industry will be working on you in every way to make you part of the problem, whether you see it or not. If you don't have the strength that is needed, get out. If you can't do either, you are a slave in a death machine. I'm sorry to put this so harshly, but that's the way it is.

Truth in the Other Extractive Industries

In the short term, the clean energy transition requires an expansion in extraction of certain minerals, especially for batteries. Overall, however, the quantities of extracted materials will need to decrease by a long way. The world will be using fewer materials, and more of these will have been recycled. A mining company that doesn't face this reality is deceiving itself and the world. Everyone in this industry needs to be talking about the need to extract less and to do so with more environmental and social care.

Truth in the Aviation Industry

However you look at it, we need some flying in the world. There is a case for some business flying, some love miles to see relatives and friends, some leisure flights and some flights to help the people of the world to understand each other better. But at the same time there is no getting around the huge carbon footprint of planes and the gaping technology gap that lies between where we are now and low-carbon flying: in fact, aviation is probably the hardest technological nut to crack in the clean energy transition. There is a pathway for a smaller but thriving aviation industry, but it must face some difficult facts and avoid some bogus narratives. As discussed in Chapter 3, there is no such thing as 'sustainable aviation fuel'. Unnecessary flights cannot be 'offset' (except, arguably, through direct air carbon capture and storage, at around $1,000 a tonne, meaning a cost of around $3,500 for a return flight from London to New York).[26] Flying is a very high-energy activity. There are no foreseeable technologies for low-energy flying, and we have already seen that the world needs to cut its energy demand. So, the aviation industry needs to face the reality of its impact and allow its customers to do so honestly. You may think your flight is worth it. I sometimes tell myself I can justify a flight for business reasons, because I am doing it to advocate for the huge systemic changes that we need. I have even taken the odd short personal flight in recent years – but you should not try to hide from the reality of the footprint, and the airlines should not encourage us to do so. If you think you can justify it, then say so, but don't pretend it doesn't have an impact.

If you work in this industry, please help it to be honest, and do not push for expansion.

Sustainability for the aviation industry involves some contraction. Anything else is a lie.

Truth in the Food and Farming Industries

As we have seen, the science is calling for radical changes in what we eat and how it is produced. The world's farmers have kept us alive and, in the UK, stopped us starving through two world wars. Many farms are the way they are because of decades of government incentives and science that failed to understand the impacts on climate and nature. Now, our ask of the farming community is greater than ever: not just to keep us fed, but to do so while looking after climate and nature. Agricultural science is incredibly complex, but while there is still a lot we don't understand, some of the big changes we need in our food system are crystal clear. To achieve them, we are going to need more farmers than we have now, and they are going to need more support, both financially and in the development of new areas of expertise. At its best, it must be one of the world's most fascinating, meaningful and rewarding careers. But right now, it is very challenging or even impossible for many farmers to earn a living sustainably. If this is the case for you, I think there are three imperatives. Firstly, do the best you can. Secondly, keep a clear view of the best available evidence, even if you don't feel able to farm in the way it implies. Don't be tempted to go along with bogus science because its messages fit better with your existing farming model. Thirdly, please campaign for the support that you need in order to be able to farm in the way that you would like to.

The food processing and retail industries need to do a much better job of letting customers know what lies behind the food they supply, without greenwashing or rose-tinting it. If the reality of the chicken on sale is an antibiotic-packed, bird-flu-inducing hideous experience for the chickens, then don't pretend otherwise. The real solution, of course, is to sell something about which you can tell an authentic, positive story. Most of the global meat, dairy and fish industries rely on customers not understanding, or not thinking about, the supply chains when they go to the shops: the desperate living conditions, the antibiotics, the sea lice, the pandemic threats; to say nothing of the climate impacts and the pressures put on both nature and the global food supply.

The Future of AI and the Tech Industry

We saw in Chapter 2, 'Standing Further Back', that efficiency gains do not by default bring reductions in impacts, because we end up doing and having more by a larger amount than is justified by the efficiency gain, which backfires in terms of environmental benefits. And we saw in Chapter 4, 'The Middle Layer of the Polycrisis', that humanity needs a new relationship with technology: one in which we are in control, instead of being slaves to adopting it every time it offers an efficiency improvement or a market opportunity. This is a massive challenge. AI, perhaps more than any other technology, is accelerating with very little human agency over the trajectory. Just as with the fossil fuel industry, there are fortunes to be made, and therefore powerful people and well-resourced companies resist any form of constraint.

AI can be used to reveal the truth, but it can be used even more easily to create bullshit and increasingly to persuade people of almost anything at all.

It is not sufficient for AI to have good applications. To be responsible, any technology business needs to be able to show that it will be used for *net* good; that the good will exceed the bad. That won't happen by accident.

Truth in the Gambling Industry

Why do I pick on this industry? Because it is an exemplar of social destructiveness and, as such, can only continue by being dishonest with the world. We've seen that dealing with inequality is a key element of both the climate and the wider Anthropocene challenges. The sad truth is that this industry makes its profits, not out of those having a modest bit of entertainment (as gambling adverts would have you believe), but by tearing apart the lives of addicts, often taking everything they have.[27] Britain's highest-paid woman, the billionaire Denise Coates, who founded and majority-owns the online gambling company 'Bet365', was awarded a CBE under David Cameron's government for 'services to the community'. Whatever the social impact of Bet365, her personal reputation is arguably enhanced in some people's eyes through the Denise Coates Foundation, which in 2022 (according to a report in the *Guardian*) was sitting on over half a billion pounds in reserves while paying out just £6 million to 'good causes'.[28]

So, if you are in this industry, get out. And if you know someone who is in it, don't let them think that their job is socially acceptable. The test of whether gambling is OK is whether someone is syphoning off profits from it. So, the charity raffle is OK, as is the sweepstake at work, as long as everyone can happily afford to lose their stake and no one is treating it as an income stream or taking a cut from running the scheme.

***

I've only listed a few examples of industries, but could easily have written plenty about consumer goods, pharmaceuticals, construction and housing, defence, research, the automotive industry, education and many more. Every industry in fact. The core questions for each industry remain the same: is it benefiting humanity as a whole, and is it helping to restore the planet?

What Does Greenwashing Look Like and How Do We Keep It in Check?

Greenwashing is easy. Most companies can make themselves look good to most of the public most of the time. It can be very tricky for them to lie, but they don't have to. Greenwashing is the ever-more-sophisticated artform of making a company look as though it is respecting the environment more than it does. It is the business version of bullshit, and as such takes place with varying levels of consciousness. Very often, greenwash is not so much a case of deliberate deceit, but more a result of staff failing to ask deep and searching questions: everyone's lives are more comfortable in the short term if no one looks too carefully under the surface.

Here are some examples, with the type of deception, as described in the Taxonomy of Deceit in Appendix 1, in brackets:

- In its original planning application, West Cumbria Mining talked a lot about local coal having lower transport emissions than imported Australian coal, but omitted to mention the emissions from actually burning the coal in the first place. (Misdirection of attention)
- Airlines offer customers 'offsets' for insanely cheap prices, creating the impression that the flight is environmentally friendly. They do it because they know it will make you more likely to fly. (False impression)
- A company calls itself 'carbon neutral' or 'net zero', creating the image that there are no climate impacts from its activities, even though the calculations do not take account of the carbon footprint of anything the company buys, or the emissions resulting from the use of its products. (False impression)
- The chief executive of an oil company produces a video about their renewables investments, with a wind turbine reflected in his eye, even though he knows the level of investment in new oil and gas exploration is far higher.[29] (Misdirection of attention)
- Coca-Cola, which emerges from an audit by Break Free From Plastic as the world's number one plastic polluter, advertises itself as eco-friendly and sustainable.[30] (Lies)
- An electric car company says its car has no emissions, even when there are emissions associated with generating the electricity to run it and building the car in the first place. (Loopholing)

Details of the nine companies picking up the biggest greenwashing fines worldwide so far are in this endnote, with Volkswagen famously top of the list.[31]

What If I Sniff Out Dishonesty in My Industry?

Our sense of tribe has to be global, not centred on the people we have drinks with or go to work with.

A passenger on the London tube will hear over and over again the message 'If you see something that doesn't look right, report it to a member of staff: See it, Say it, Sorted.' It is their attempt to make it normal to call out what isn't right, even though it is uncomfortable. We need the same in the workplace, although it may sometimes entail reporting it outside the organisation. To make this easier, we need to greatly tighten up the rules on whistle-blowing so that people can call out wrongdoing without paying too high a price, but in the meantime, it still needs calling out.

Stepping outside the company groupthink takes real independence of mind and courage. Usually, the company culture will have a storyline that justifies its behaviour. You need great critical-thinking skills to see past this and then courage to do something about it. You might, for example, be told that you are naive; that you just don't understand the way the real world works. Whatever the line, you will probably be under massive psychological pressure, and you will almost certainly end up questioning yourself. You will be encouraged, though not necessarily explicitly, to distract yourself and move on.

I'm not just talking about calling out the Post Office and Partygate-style dishonesty, which so many people had a duty to blow the whistle on at the time. I am also talking about challenging cosy comfortable narratives that companies have for themselves that justify how or even *why* they operate: for example, why they extract excessive materials, create products that are not needed, perhaps even encourage people to do things that will leave them poorer, unhappier and unhealthier, and why they invest in companies that do bad things.

If you see show-stopping problems in the company you work for, and you can't change them from within, then, now that you've seen it, you have to get out if you possibly can. And if you feel that you really can't, then you are a bonded labourer. If you *can* get out, you won't look back.

Are There Any Good Businesses Out There?

Plenty. Admittedly most are a conflicted mix, sitting somewhere on a spectrum between wonderful and appalling. Most are trying to make themselves look a bit or a lot better than they really are. Most are avoiding – to a greater or lesser extent – confronting the most uncomfortable questions, and doing so with varying degrees of consciousness. But while there are some awful companies wreaking such havoc that their senior managers are probably more deserving of jail than most of the inmates of our over-crowded prisons, there are even more examples of truly exemplary and Anthropocene-fit companies out there too.

I can't give anything like a definitive list of the best ones. But here, by way of illustration – and with an apology to the many fantastic companies that I am missing out – are just a few that have come across my radar, some of which I have even been lucky enough to have worked with.

On the UK high street, Timpson is championing the circular economy through its repair services while simultaneously being a great employer and a force for social good. I spent just one day with the retail side of the Salvation Army, but it stands out perhaps more than any company I can think of for the consistency with which it applies its positive values to absolutely everything it thinks, says and does.

Patagonia, of course, has pioneered and trailblazed sustainability in the clothing industry, with its free repairs, longevity and sustainable sourcing, and its social and environmental campaigns.

Triodos Bank and the Ecology Building Society are trailblazers in personal finance. In asset management, LeapFrog is all about social and environmental change for developing countries, while the Global Returns Project is not about financial return at all, but helps people to give money away more effectively.

In the energy sector, Ecotricity and Good Energy have both championed renewable energy for decades, and now Octopus is also doing great things to enable green homes, renewable supply and a balancing of the UK's electricity grid.

I am approached by entrepreneurial start-ups with high ideals in many sectors. We work with a few of them, and some of them go on to make a real difference. Local to me, there are farmers doing amazing things, an organic grower that simultaneously provides mental health services to volunteers, a zero-packaging food retailer, electric bike hire start-ups ... I could go on and on, picking examples from most industries.

I feel sure that positive-impact businesses greatly outnumber the monsters. The problem is that very often they are small-scale, whereas it only takes a handful of terrible huge corporations to be destroying the world. How can we change that balance? Every time any one of us spends money, either in our personal lives or through our work, we decide what type of businesses we want to support. If you are lucky enough to have some choice in who you work for, you are choosing what kind of businesses you lend your talents to. I'm often quite sceptical of company certification schemes, but B Corp certification does mean something for anyone looking for ethical suppliers.

10 THE EVOLUTIONARY CHALLENGE AND WHERE EACH OF US FITS IN

Now let's step back again from the practicalities of getting more honesty into our politics, media and businesses, to remember that all this is the pragmatic edge of the huge evolutionary challenge for humanity to rewire itself, such that we can survive and thrive in the Anthropocene that we have created. A massive collective wake-up is needed, without waiting until it is too late, followed by a significant cultural leap for humankind. What we are talking about here is far more significant than someone stepping onto the surface of the Moon. We are talking about a major development in our psychological and social make-up.

These are surreal times, with the clouds building, and the absurd incongruence between the messages from science and the political and social narratives. On one level, these times are still so very comfortable for relatively well-off people like me, even though they are already so uncomfortable for many others. There is almost certainly – but not inevitably – a huge storm ahead. The values we promote now and the actions we take can change the ferocity of that storm, as well as the impact that it has on human lives.

Even without the Polycrisis, the way the global system functions is clearly such a sub-optimised way of running society. In previous decades, it represented a lost opportunity to live better: now it threatens the survival of billions.[1]

10 The Evolutionary Challenge

The choice is radical change or untold suffering and death. We don't quite know when, but we can all see the global society and economy getting wobblier, just as self-correcting complex systems tend to do before they crash. While many of those who care most deeply about the Polycrisis have been concentrating mostly on the outer layer, the crust of the problem, many are now turning their attention to the middle layer and the core. Of those who concentrate on the outer layers, I know heroic people who seem to not expect to get anywhere, but carry on because at least sounding the alarm feels better than lying on the beach and letting it all fall apart. I even know one or two who have died looking back on their life's work as a frustrating failure. I don't want you or me to end up like that. I still think the possibility of being able, broadly speaking, to catch things in time to avoid catastrophe are well worth fighting for. And even if we are too late to stop things getting awful, we will still be incredibly grateful for anything we managed to do before the trouble escalated, as it will stand us in better shape to head off the worst. To do any of this, we must be clear about what we are facing.

In this context, more people, myself included, are asking how best to think about it all: how they should live, what part they should play, and, having chosen not to ignore it, how they can do all this without going mad. That is what this final chapter is about. Those who want us to carry on burning fossil fuel would love us to flip from denying the problem to hopelessness (from the left-hand end of the Kübler-Ross curve to the depression dip) and then stay there without ever taking action, or else believe that we will be fine because we always have been – which could be claimed by any civilisation or species throughout history until it saw that it was dying, and which fails to acknowledge the uniqueness of the Anthropocene challenge.

Increasingly, I am finding that those who have thought about the state of the world the most carefully are taking what I would describe as an extremely big perspective. Friends and colleagues are saying things like 'The planet will be fine. It is just the people that won't be.' Or 'Plenty of species go extinct. What is so sad about us joining them?' I too find solace in standing further back from time to time, not because it makes me want to give up but because perspective eases the discomfort. In Buddhist philosophy, as well as the whole of existence being interconnected (which has also been a running theme in this book), everything is impermanent, so it's a good idea not to get attached to the way things are. In our culture, we give ourselves so much importance. We talk as if we are the best species, but there is at least as strong an argument that we are the worst.

When Lord Martin Rees, the Astronomer Royal, opened the building that my office is in, he reminded us all that the Earth will be absorbed by the Sun at some point, even though by the time that happens, he said, the beings who inhabit the Earth will be as different from us as we are from bacteria. That is quite something to contemplate. But it is clear that humans face existential challenges in the much nearer term than that. As I see it, it would be nice to have at least a few more centuries. And if we can achieve that, maybe we really will be able to go off to Planets B, C, D and E. And then, those who still want to may even be able to resume the traditional approach to growth which is so problematic right now. Having managed the evolutionary challenge of becoming capable of living well together on Planet A, I might even trust us to go exploring without mucking up everywhere else we visit. But this speculation is a distraction from the immediate practical challenge, of which space travel is not a part.

Before any of us can decide how we might best be of help, we need to get clear about whether there is any point in trying, or whether our fate is already sealed.

Have We Left It Too Late?

We don't know.

None of us knows for sure how much trouble we are heading for or how fast, but reflecting as I've been writing, this book has been a journey for me too. I don't have a crystal ball, but overall I'd guess that the next few decades are likely to get uncomfortable for humanity. The faster we wake up, the better we can mitigate this.

> *However well or badly things turn out, society is going to need all the honesty, kindness and joined-up thinking that it can get.*

What Makes Me Pessimistic?

The severity of the crisis and the stubborn inadequacy of our response so far. The global systemic nature of the challenge, alongside the pitiful lack of international cooperation. The unhelpful and deluded narratives of progress. Our tendency to select the leaders who don't have the qualities that we need.

My pessimism comes from knowing that decades after the publication of *Silent Spring* and *The Limits to Growth*, and despite absolute clarity from the scientific community, and even after 29 climate COPs, we are *still* sleepwalking towards disaster. We carry on accelerating our greenhouse gas emissions and our wider destruction of the natural world despite the evidence that is now *screaming* at us to change tack. The resilience of human sleepiness to the dangers we are facing, and to the depth of change that is needed beyond superficial techno sticking plasters, is truly remarkable given the weight of the evidence. The domains of politics, media and business are pulling each other back from the roles they each need to play in sorting things out, when they should be pushing each other forwards. The almost-free market is so dominant that we have inadequate defence against the negative impacts of any industry that sees a commercial opportunity. We continue to allow charismatic but dishonest mischief-makers into positions of power. Too many of us – including a majority of US voters – vote for people who don't care about us or the wider world, and read 'news' from sources that obviously can't be trusted. International affairs are still dominated by wars, power games and rivalries, even though it is imperative that we all get on the same side. If an alien from another part of the Universe arrived and turned on the radio to hear about what is happening on Planet Earth, it would take them a long time to realise we have a Polycrisis going on in the background of our day-to-day squabbles.

What Makes Me Optimistic?

Despite the appalling US mess, the world as a whole might be very close to the huge social tipping point that will make all the difference. Maybe it just needs another big push and a splash of courage. Most people want action, but don't realise how many others want it too, and are waiting for others to break ranks first.

Social tipping points can happen fast. We might be on the cusp of one right now. We might not need an unimaginably traumatic event to shake us into action. The signs that a complex and self-regulating system is getting ready to change can be hard to see, but the social pressure for change is growing. The complacent stuck-ness that has plagued us for decades is beginning to wobble, just as we would expect in the run-up to an abrupt change. In some countries, we might be almost there, and if we push hard now, we might be in time, and such change might find a way of spreading globally. On both sides of the Atlantic, we've seen some desperately inadequate people voted into power over recent years. In the UK, the consequences of this are becoming so plain that more of us may be starting to learn the lessons from our experiment with truthlessness, perhaps triggering a wider reflection on what leadership needs to look like in the Anthropocene. Meanwhile, with the re-election of Trump, the US has taken its experiment with post-truth to a new extreme. My fear is that the rest of the world will feel the consequences of this dreadful act of collective destruction well before they play out directly in the US, and even then, they may not be widely acknowledged for what they are.

In 2024, a survey of 130,000 people in 125 countries found that 89% want their governments to do more to support climate action. Eighty-six per cent supported 'pro-climate social norms', and 69% were willing to contribute 1% of their income to addressing climate change. (Which is impressive, bearing in mind that many of the remainder are living towards the bottom of a deeply unequal society and have zero disposable income.) These are all very high percentages, from a very large global survey, and the analysis was published in *Nature Climate Change*, a highly respected, peer-reviewed journal. In other words, it constitutes strong evidence that humanity wants action on climate.[2] So why aren't we taking it yet, and what will it take for us to tip?

Importantly, the survey also found that in every single country, people seriously underestimated everyone *else's* willingness to act. Globally, while 69% were willing to sacrifice some of their income, on average people thought that only 43% of their fellow citizens would be willing to do so. The willingness to act was significantly underestimated by those surveyed in every single one of the 125 countries (Figure 40).

This data chimes absolutely with my own anecdotal evidence. I give a lot of talks to many different audiences. Some are festival-goers, but most are in business settings in which I have no reason to think that the people I'm with are unrepresentatively green-minded. Recently I've been

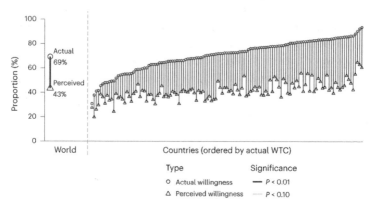

Figure 40 Willingness to contribute (WTC) to climate action by sacrificing 1% of income, compared with perceived willingness of others to act, in 125 countries. In every single country, people significantly underestimate other people's willingness to act. Across all countries, 69% are willing, but people believe the figure is only 43%.[3]

trying a little experiment. I have run it about 25 times now and with several thousand people. Firstly, I spend 10 minutes or so presenting a summary of the Polycrisis and our progress in dealing with it. Exactly along the lines that I've outlined in this book, I lay out what I think is a pretty stark picture, in which we are not definitely doomed, and there is still, as far as we can tell, everything to play for. I take care not to over-blow it, but not to under-egg things either. Then I ask for a show of hands of those who think that in broad terms I've got it about right; those who think I've fundamentally exaggerated the situation; and those who think I've fundamentally underestimated the seriousness. I do my best to encourage anyone who thinks I've exaggerated to be honest and brave enough to say so. Sometimes I do this voting online, so that it is a secret ballot. Whatever the audience, the results are always striking. The vast majority (75 to 95 per cent) always think I've got it about right. There are always more people who think I've under-sold the seriousness than those who think I've over-played it, so every time the centre of gravity of the room thinks the situation is slightly worse than the one I've described. I get everyone to look around and see the hands in the air. It can be an illuminating and critical moment. People suddenly see that they are far from being in a minority that understands the incredible seriousness of the situation, but that nearly everyone else gets it too. If they all work for the same company I can then say, 'Right! Now you all know you have both permission and responsibility to make sure that everything that is said in every meeting is coherent with this shared understanding, and to challenge anything that is not.'

The difference between how people think and how they think others are thinking is hugely important. Casting my mind back to the COVID-19 pandemic, I remember how uncomfortable it felt either to be the only one in a room wearing a mask or the only one not wearing one, regardless of which behaviour made the most sense. So, turning to the Polycrisis, or at least to the climate element of it, most people want to see strong action, but because most of us find it so desperately uncomfortable to break social norms, we are still all looking around to check what everyone else is doing, and there are very few who dare to make the first move. It is so embarrassing to be the only one in the room who is calling for action that is genuinely commensurate with the situation we are in, because it is such an unusual thing to do. Virtually all the social messages we pick up from our media, friends and relatives tell us to carry on as if the climate emergency either doesn't exist or only requires a superficial response. But gradually, more of us are realising that most others want change too and might even be ready to break ranks to get it.

Are Humans Too Fundamentally Selfish to Survive?

No. We don't have to be.

One of the many unhelpful fake narratives that has been doing the rounds is the idea that we are all just in it for ourselves. Those who are living the most selfish lives love the idea that everyone is like this because it makes them no worse than anyone else and justifies how they are living. And it is true that we have struggled to contain our psychopaths and have given too much power to people who lack both honesty and empathy. But most of us are not like this, and a disastrously selfish culture doesn't need to dominate, if we don't let it. The evidence from neuroscience, from history and from psychology, as well as from traditional wisdoms, is that we are capable of cultivating values of mutual respect when we need it in order to survive.[4] It turns out that, contrary to the suggestion of Richard Dawkin's seminal book *The Selfish Gene*, natural selection of DNA favours those that collaborate the best, not the most selfish genes. It also turns out that, contrary to the fictional story in *Lord of the Flies*,[5] when a group of school kids actually did get shipwrecked on an island, they lived well together and were in great shape when they were rescued, remaining friends into their nineties. It furthermore turns out that several famous research experiments that supposedly prove our inherent selfishness were rigged by the researchers in order to deliver these results. The famous Stanford Prison Experiment had to be restarted because first time around the prisoners and the guards got on too well and set up a happy commune together, frustrating the researcher who wanted to write a paper with a different finding. (Rutger Bregman writes wonderfully about the examples above and more in *Humankind* – which I recommend for anyone who wants to feel better about humanity and its potential.[6]) The evidence is that we are capable of fostering different cultural values, and these can have dramatically different consequences, from genocides and world wars to peace, harmony and respect. To the extent that we can influence anything at all in our lives through human willpower, we can choose the values that we operate from and those that we cultivate in society.[7] What we haven't succeeded in doing yet, but have never previously needed to, is cultivate the value of global universal respect. Now is the time, because we can no longer get by without it.

We are not inevitably too selfish to survive.

How Should We Deal with Crisis Anxiety?

Face the situation and take action.

The underlying knowledge that we are heading for a crisis and yet carrying on as usual has to be deeply unsettling for all of us. It hangs like a lead weight on our collective mental health. This must be especially acute for our kids, who see the world with fresh eyes, and have a greater ability than adults to spot when an emperor isn't wearing any clothes. It is mentally draining to be asked to live in one way by society, while seeing that this is a disastrous way to live. We can try pushing the problem to the backs of our minds, but the trouble is, we know that is what we have done. The knowledge of this nags at us, even if we aren't consciously aware this is happening.

One response is to deny the problem, not necessarily completely, but to deny its depth; to pretend that technical fixes are all that we need. I've written that I think this does more harm than good. As Kübler-Ross noted, we can't start to make the most of our situation until we have a deep and realistic acceptance of what that situation is. Another response is to give up and say we are doomed; that also isn't good and isn't in line with the evidence.

So, while I don't claim to have the perfect answer to climate and ecological anxiety, here are a few things that I think help:

- Firstly, in terms of how we look at the problem, the most psychologically healthy approach is to look the whole problem in the eye: to do our best to understand the reality of the situation in all its dimensions and depths as best we can, neither exaggerating nor trivialising. As I've tried to do in this book, we need to stand back for perspective, then burrow under the surface and also join it all up. To be healthy, we need to allow into our consciousness the most complete, mature and nuanced awareness of what is going on that we can get.
- We should then say it as we see it and without compromise. Mental health is helped by congruence between what we say and what we think. This is sometimes going to mean that what you say is uncomfortable in the environment that you are in.
- Take action in whatever ways each of us thinks can have the greatest agency. I hope by the end of this book you will have increased your sense of how to do that. Find some stuff from the checklist, or elsewhere, and get going on it. It feels good to be doing the right thing, whatever the state of the mess.
- A big philosophical perspective might help. We all have different life philosophies, and I don't want to tell you what yours should be, but maybe it is helpful to remember from time to time that we are infinitesimal specks in the Universe. We humans often talk about ourselves as if we are more special and important than perhaps we are. Other species often go extinct, usually because of us, and in the very worst scenario, maybe we are the one that most deserves to do so. The Earth is just

a minute part of even the known bit of the Universe. Individually, each of us is going to die anyway, and very few of us know when that will be. But it would be nice if humans on Earth could have a few more decades, centuries or even millennia.
- None of us need to be alone with this. It is easier to face problems together, so try to find a group, a volunteer organisation or other community of people who share your perspective. Share with friends and family if you can.
- Finally, try to enjoy life, and the people and activities that you love. Regularly do things to recharge. To put it in a slightly trite way, there is no point saving a world that is no fun to live on.

When the Challenge Is So Global, What's the Best We Can Hope For?

We've seen that the problem is global and systemic, and it is important to understand that local actions, even on the scale of, for example, sorting out the politics, media and business culture in the country in which you live, don't really help unless they have *systemic* influence.

I don't want to be utopian about this. I find pie-in-the-sky dreaming almost as depressing as doom-mongering. Even if we start playing our cards much better from now, humanity might already be committed to some very nasty consequences of our blundering carelessness to date. Even the best global governance that I can dream of might not be able to head off enormous human suffering in every country of the world. And with Trump, Putin, Xi and others in power, my best conceivable scenarios feel very far off. But whatever trouble we find ourselves in, any improvement we can bring about in the quality of our governance will be invaluable to us all.

We all need a believable theory of change into which our actions fit. It needs to be radically more believable than today's central narrative of 'net zero' stuck on top of business as usual, media as usual and politics as usual.

So here is a best-case scenario that all of us can push towards. And the closer we can get to it, the better things will be, for us all:

- Starting right now, we create a culture that insists on far higher standards of honesty than humans have ever had in their governance. It starts as a galvanising concept in one country – perhaps the UK. It taps into a vein of energy. People see the benefits. So it grows. Other countries see what can be done. The standard of politicians goes up radically because the new culture improves the standards of even the worst politicians, media and businesses, and at elections, the best ones are voted in. High-quality leadership in some countries moderates the behaviours of the world's worst leaders and in time brings Anthropocene-fit leadership to the world.
- New, thoughtful conversations become possible. The quality of decision-making improves, first in some nations and then internationally. The movement is so constructive that only the sociopaths and psychopaths don't like it, and it is strong enough to keep them at bay. Humanity finds a way of getting itself out of the grip of psychopaths and into the hands of people who care about others more than their own careers.
- Parallel shifts take place in the media and in business, and these three domains now reinforce each other in raising their games – spurred on and held to account by an increasingly informed and engaged public. Lives start to get better as a result. The evolution of human decision-making takes off, and for the first time we begin tackling the Polycrisis in a way that is fit for the challenge.

How Can I Work Out What Part to Play?

Be strategic. Think big. Find a fit between what the world needs and what you might be capable of offering.

Start with a big question:

> **How can I help to create the conditions under which humanity can undergo the multi-faceted systemic change that we so urgently need?**

To answer it, here is a very simple and well-established three-stage strategic process:[8]

(1) Think about the global situation. What's the big picture? What does the world need right now?
I hope this book has helped with that, but please feel very free to add to and tweak my analysis in any way you like.

(2) Think about the kind of things you feel capable of offering or becoming able to offer, given your own skills, resources and your potential to develop new skills.

(3) Find a way of fitting the two together, so that you deploy your capabilities to maximum effect to meet the needs of the world (Figure 41). Don't worry if you are not sure you can succeed.

This is the big empowering way of thinking about what you can do that gives each of us the chance to have proper agency, even on a problem that is so huge and global that it is easy for each of us to feel like an insignificant powerless speck. While none of us will single-handedly be able to change everything, **we can each be a meaningful part of the change, and that is enough**.

So where does that initial big question take us to? The answer is different for each of us, but it opens up a wealth of possibilities. It took Greta Thunberg to sitting outside the Swedish Parliament and then on to confronting world

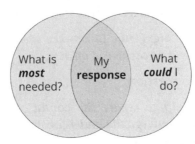

Figure 41 Be strategic. Work out what is going on and what is most needed. Work out what you can do or could become able to do. Find a fit between the two.

leaders. It took Roger Hallam, Gail Bradbrook and eight other activists to founding Extinction Rebellion. It took Adam McKay to directing the film *Don't Look Up*. It has taken me to writing books (even though I was usually bottom of my class in English at school), to running a business (even though I never really think of myself as a businessman) and to doing research and analysis when I see an important gap that I think I can help to fill. I also get involved with a few other initiatives that I have identified as having potential for high leverage on the system where I might be able to contribute. It has even made me take to the streets from time to time to lend one more hand to something important and inspirational. I didn't start out as an author, a businessman or an academic. I started out trying to fill gaps.

That's me, given who I am. But we are all different. For some, it means going into politics. In my work I meet so many amazing people who have dedicated their lives to a better world in different ways: journalism, campaigning, research, teaching, farming, sustainable entrepreneurship, environmental law, new types of investing and more. I know people who have been prepared to go to prison for actions that they don't even know for sure will do more good than harm. And there are others whose contribution is to simply provide infectious role models of honesty and global kindness. We are all different. My point is to be strategic, so that you maximise your leverage and your impact.

Alongside these big decisions, which might transform your career and your pastimes, are the high-leverage everyday actions about *how* you do what you do: how you go about your work, obtain news, spend money, talk to your friends and family, vote and so on.

The checklist below is a smattering of ideas, big and small. Please take what you can from them and add your own.

What Can I Do? A Checklist of Actions

This list is for people who recognise that urgent global systemic change is required, who see that we are not yet getting anywhere and who want to have some agency in that context.

Politics
- Ask very carefully who you can trust and how you know that. Use the criteria in this book or make up your own.
- Talk to your friends and family about the importance of honesty and empathy in politics. Perhaps also try to help people see why voting for narrow self-interest is self-defeating.
- Challenge those who don't prioritise these characteristics for evaluating politicians or have poor criteria – even friends and family.
- Make sure your electoral candidates know what you are looking for before the election.
- If you get a chance to vote, choose the best candidate who stands a realistic chance, and remember that even if all candidates dishearten you, there is a difference between them and by voting you raise their game.

- Outside of elections, if you get an opportunity to be involved as a citizen in the democratic process, take it.
- Consider getting involved yourself, either as a politician or as an activist. Consider joining or working for an organisation that pushes for a higher standard in politics.

Media
- Make a very careful choice about which media you can trust and why. Use the criteria in this book, or apply your own, as long as they are at least as good as mine. Try to have a variety of sources.
- If you can, help to fund the media that you want to support. Good media needs serious funding, and if you want it to raise that money without needing to compromise its content or promote unsustainable businesses through its advertising, then, if you can afford to, consider paying a lot more than, say, the price of a standard daily print paper. If you are in a position to donate, consider good media as one of the highest-leverage causes you could support. (See Appendix 4: Resources.)
- Take issue with friends, family and colleagues who are undiscerning with their media or buy it because they like the TV guide in it. Help to make it culturally unacceptable to consume media that is careless with truth or disrespectful.
- Consume social media with care, choosing who you follow based on information from outside that social media platform. Make sure your own content is in line with the world you want to see.
- Consider joining the media to be part of their reformation. Maybe start a podcast. Perhaps write, either for the national or the underfunded local press that desperately needs good content. Maybe even retrain to join this profession.

10 The Evolutionary Challenge

Work

- If you possibly can, work for an organisation that is part of enabling a more sustainable future. Ask carefully whether the goods and services it provides, and the way it does so, are helping to enable such a future. Apply very careful criteria, such as the ones I outline in this book, or make up your own, but dig beneath the surface and don't buy the superficial storylines. If you feel you can't leave an organisation that you don't like, then recognise that you are a bonded labourer and try to engineer your freedom. If possible, consider reducing your personal financial requirements in order to gain freedom.
- Help your organisation to become better, by challenging it and supporting it across everything it does, from its overall strategy right down to everyday decisions and actions. Help it to ask the big questions about why it exists, and how it helps the world to make the changes it needs to. Do not settle for superficial responses. If trying to change from within, keep asking yourself carefully whether you are helping, or becoming part of the problem, and try to have plenty of contact with people who are outside your organisation's mindset.
- Challenge friends and family who work for organisations that don't meet your criteria. Similarly, for those reading poisonous media, it needs to become socially unacceptable to be part of a people- and planet-wrecking business (unless you really are transforming from within, which is very tough).
- Push for responsible buying decisions, remembering that everything your organisation purchases supports one business or another, and therefore pushes for one type of future or another.

Activism, Volunteering and Giving

- Outside of any paid work, consider giving money and/or time to high-leverage initiatives, especially in politics, media, upholding of truth, ethical standards and universal respect for all people (including peace).
- Do not think you can 'offset' a destructive job with philanthropy.
- Think carefully about getting involved in campaigns, and if so, which ones and how. My own view is that all protest should not just be non-violent, but should send out an overall positive message, even when some disruption is involved. I've written elsewhere about what I think effective protest looks like.[9] I don't think anyone knows for sure what approach is most effective. To me it is very clear that, at their best, Greta Thunberg and Extinction Rebellion, for example, have been huge positive influences on the world. I find it hard to know whether more disruptive actions such as Just Stop Oil's did more good or more harm, but either way I respect the intention and the courage behind them, and I resist simplistic positions on their impacts. We do, after all, have to find a way of being more effective than anything anyone has tried so far. All of us should carefully consider activism.

Sustainable Personal Lifestyle

- Living more sustainably deserves a place on the checklist partly because it brings integrity to everything else you do. It will leave you feeling better about your life. And by making improvements personally you will come to understand the global transformation better, since your own micro-experience mirrors the global challenge in so many respects. Everything you buy, or don't buy, invest in or divest from, pushes for one future or another.

10 The Evolutionary Challenge

- Do not beat yourself up for your shortcomings, but don't let yourself off the hook either.
- Don't bother with carbon offsets. Face up to your impacts, and if you have any spare money, donate it to the causes with the highest leverage that you can find: good media, good campaigning or some other cause. If you can afford to give money away, put it towards the most impactful cause you can find.
- If you can, donate money and/or time. Both are good things to do. It is better to earn your living in a way that genuinely helps people and planet, and not have spare time or cash to donate, than to earn a higher salary doing a job that is fundamentally destructive and donate the spare cash philanthropically. I am sceptical of the philanthropic model whereby the over-rich promote themselves as saints for donating a small proportion of a wealth that may have been earned in ways that don't look great if you ask searching questions. With money especially, be strategic. Pick the cause with the highest leverage and put all your money into that. There is little point spreading your funds around unless you are such a big donor that your contribution alone is enough to saturate the organisation that you identify as the most beneficial. When it comes to putting in your time, it is a slightly different equation because you are not just looking for a good cause, you are looking for one that you can support given your own particular skills, and if possible something you will also enjoy.
- Prepare for unstable times. Given where we are and where we are currently heading, it is wise to think a bit about readiness for an unstable future. I don't mean building a self-sufficient concrete bunker in your back garden. There is probably some value, however, in all of us making sure that, in the worst scenario, we can purify our water for a few days and eat, just because

this is easy to sort out and it would significantly soften the blow if the system wobbled badly for any reason at all. It could help to contain a situation that could otherwise break down completely. But the biggest precautions we can take are to have a resourceful mindset and a strong sense of community.

Look After Yourself
- Cultivate thinking skills, values and inner development goals. I have no claim to special expertise here, but find a source of inspiration or make it up as you go along.
- Enjoy and appreciate life where you can, because there is no point saving a world that's no fun to live in.
- Be close to family, friends and community if you can, because these are great when times are good, and essential when times are tough. Develop flexibility and resilience, because business won't carry on as usual. (A note to billionaires, by the way: these assets will be far more valuable than a multi-million-dollar fortress on an island.)

Finally ...

I have tried to write a practical book about how we can best approach the biggest problems we face. I know there is plenty more detail that could be added or tweaked, and you might think there are bits I've got wrong. But the further I've got into the writing, the more confident I've become of the central proposition: that those of us who care about the physical realities of the Polycrisis need to focus most of all on enabling a new quality of decision-making, which is essential if we are to live well, here in the Anthropocene.

The polycrisis is symptomatic of a *metacrisis*. The collection of presenting symptoms have common, deep roots that must be addressed if we are going to get anywhere at all. Sticking plasters on their own haven't yet helped and nor will they do so in the future. So I have tried, as promised, to stand further back, to dig deeper under the surface and to join up the many dimensions of the challenge. I have started from the biggest macro view and moved through to the practical implications for how each of us might think about how we live, the actions we take, and how our society thinks and functions. I have tried to drill far enough down through the layers of this metacrisis to identify the most critical practical levers that each of us can pull. I hope my efforts are enough to be of practical use, even though I have stayed well clear of the deeper philosophy that some can persuasively argue is also required. That is not because I disagree with them, but because philosophy isn't my field, and there is only so much one book can take on.

And I have tried, not perfectly, to apply some of the thinking skills that it is clear we all need to get better at, along with some simple values I think we all need to cultivate, both as individuals and collectively.

Do let me know (respectfully please) what you would add, tweak or delete from this book: mike@climateoftruth.co.uk; and do check out the website www.climateoftruth.co.uk.

Thank you very much for reading.

APPENDIX 1 A TAXONOMY OF DECEIT

A climate of truth means a lot more than 'No Lies'. It means we require people and organisations to help us understand what is what, and not to mislead us in any way. So, this taxonomy is to flesh out, with descriptions and examples, a more complete range of totally unacceptable behaviours. Any other ways, not covered here, in which politicians, media or businesses can mislead us are, of course, just as unacceptable.

Lies

The most basic and traditional form of deceit, but not always the most damaging. For something to be a lie it needs to be clear-cut untrue, and the person who says it needs to know that it is false. Lies are hard to prove, and the penalty for falsely accusing someone can be high. Often, it is as far as we can get to show that something is untrue and to draw the conclusion that the propagator is *either* incompetent or dishonest. Either way, the source is untrustworthy.

To understand the seriousness of lies, and their capacity to escalate if the culprits are not held to account, look no further than these first two examples.

Example 1 Trump's Election Claim

Donald Trump's claim that he won the 2020 presidential election resulted in the storming of the Capitol Building, almost completely toppling American democracy, such as it is.[1]

A Taxonomy of Deceit

> **Example 2 Putin's War in Ukraine**
>
> In the months before the invasion, Vladimir Putin lied that he had no intention to invade, and he has told many more lies since.[2] He has imposed 25-year jail sentences in brutal conditions on Russians who have been brave enough to speak out against the war.[3]

Closer to home for me, the following examples are less dramatic but are the thin end of a very wide wedge. I could list many examples of blatant lies, but I don't need to, because the point is that you only have to identify one definite lie on a policy issue to establish that the perpetrator is totally unfit for a job in politics, *and* that any colleagues who fail to call them out if they know about it are also unfit. I write this from a carefully non-partisan perspective. Shockingly, however, the examples that follow, and the lack of challenge from the UK's Conservative MPs at the time, are all the information you need in order to know that not a single member of the early 2024 UK government was fit for office. Later in this Taxonomy we will find some evidence of dishonesty from their political rivals too, but at the moment, at least, it is much less prevalent. If you want to feast your eyes on a long list of MP lies, the highly respected former *Spectator* and *Telegraph* journalist Peter Oborne has painstakingly catalogued hundreds of them emanating from the UK government between 2021 and 2023. Go to www.political-lies.co.uk.[4]

> **Example 3 Grant Shapps's Voting Record**
>
> On 23 June 2023, **Grant Shapps** told us on Sky News that he had voted for Brexit. In 2016, he also told Sky News that he wouldn't; and he confirmed six months after the referendum that he had voted Remain. He can't be mistaken, and he has turned down

opportunities to correct the record.[5] So, in my opinion, he has lied. What do you think? It is a relatively inconsequential example in itself, but if you agree with me, it speaks volumes about the man and his fitness for office. The principle that if you can't trust someone on one thing, you can't trust them on anything will be amply borne out through this book and elsewhere.

Example 4 Partygate

This issue caused enormous hurt across the UK, with desperate consequences for trust in government. When UK MPs and even the Prime Minister **Boris Johnson** claimed not to have been having parties together during COVID-19 social gathering restrictions, these also turned out to be provable lies.[6] A small number of those involved, including Boris, eventually lost their jobs, at least in part over Partygate.[7] (Some of the Number 10 partygoers also got fined, but only £50, while other people around the country, including some much less able to pay, were fined upwards of £10,000.[8])

Example 5 COVID-19 and UK Care Homes

On 15 May 2020, Health Secretary **Matt Hancock** told Sky News, 'Right from the start we have tried to put a protective ring around our care homes', followed by **Boris Johnson** saying, 'I can tell the House [of Commons] that the number of discharges from hospitals to care homes went down in March and April, and we had a system of testing people going into care homes.' The claim was repeated by Culture Secretary **Oliver Dowden**.[9] It was all untrue. People were arriving in care homes from hospitals in huge numbers, without tests, and many homes had no access to personal protective equipment (PPE). As it happens, a member of my family worked briefly in one of these homes, without access to PPE, and it could not have been clearer that people were dying in unnecessary numbers.

> **Example 6 CV Lies**
>
> In **Tony Blair**'s application to become a parliamentary candidate, there were definite lies in his CV.[10] Why should this have been a show-stopper for his political career? Because, for example, we needed our Prime Minister to be someone who was scrupulously careful with the truth, however inconvenient he may have found it, for example when it came to talking to the nation about evidence for weapons of mass destruction in Iraq.

> **Example 7 Grant Shapps and Energy Prices**
>
> Also mentioned earlier, in June 2023, **Grant Shapps** (again), then-Net Zero Secretary for the UK government, claimed repeatedly in tweets and media interviews that ending new oil and gas licences in the North Sea would lead to 'skyrocketing energy prices'. But UK energy prices are set by global commodity prices and are almost unaffected by the UK's 1 per cent market share – as previously acknowledged by the UK government itself. Could his comments have been just a mistake? At first sight it looks conceivable, at least, that this could simply be a grossly incompetent basic error. However, CarbonBrief clearly pointed out the inaccuracy,[11] citing robust evidence and alongside proof of a further six equally egregious falsehoods in Shapps's narrative on North Sea oil licences. The proof of lying rather than incompetence comes in Shapps's failure to correct the record on any of these seven counts of truthlessness. **Rishi Sunak** repeated the skyrocketing prices claim and also failed to correct the record.

Example 8 Rishi Sunak's Claim That Just Stop Oil Funded Keir Starmer

Another example feels important since it involves a UK Prime Minister, blatantly lying about his opposition:

> *What do the unions and Just Stop Oil have in common? They bankroll him [Sir Keir Starmer] and his party.*
>
> Rishi Sunak, February 2023

In reality, Just Stop Oil would have been in no position to fund Labour, as it was desperately trying to find enough money for its own actions. It was also almost as critical of Labour's policies on climate as it was of the Conservative Party's. This was a crystal-clear untruth.[12] And Sunak's failure to correct the record makes it a clear-cut lie.

Example 9 The Brexit Bus

> *We send the EU £350 million a week – let's fund our NHS instead.*

There were several forms of dishonesty at work on the side of the Brexit Bus, and the lie that the UK spent £350 million per week on the European Union, when the real figure was more like £250 million, was a relatively small part of the total deception. Far more serious was the simple misleading implication that the payment could firstly be saved and secondly be redirected to the NHS. Eight years after the Brexit referendum, the NHS is perhaps in the worst state it has ever been in, with 300 to 500 people a week dying just from delays in emergency care, following years of chronic underfunding and certainly no Brexit boost.[13] On the impact of this deceit, Dominic Cummings commented, 'Would we have won without £350 m/NHS? All our research and the close result strongly suggests No.'[14] Whether or not you think Brexit is a good thing, the fact is that on the admission of the leader of the Vote Leave campaign himself, that vote was won on lies.[15] And the NHS didn't get any richer as a result.[16]

Calling someone a liar should not be done lightly, but if the evidence is there, it is important to be able to do so. When Labour MP Dawn Butler stood up in the House of Commons and correctly stated that Boris Johnson had repeatedly lied, she was ordered to leave the chamber because of an unhelpful ruling that MPs are not allowed to use that term.[17] That ruling needs to be abolished, so that lying, when it does occur, can be more easily flushed out.

Probable Lies

Sometimes we don't know for sure if someone is lying.

Example 1 A Feeble Excuse

The House of Commons held a vote on whether they agreed that Boris Johnson had repeatedly misled Parliament over Partygate. Over 200 MPs found a way of not taking part. **Rishi Sunak**'s excuse was that he was at a dinner that night.[18] However, several of his colleagues who attended the same dinner managed to get away in time to vote. So did Sunak lie when he said that the dinner was his reason, rather than that the dinner provided an excuse? It is not quite clear-cut enough to call it a definite lie. To me it is a feeble excuse and a probable lie, because the motive for not voting looks very likely to be the same as the reason the other 200+ Conservative MPs found for not voting on what was obviously an incredibly important question.[19]

> ### Example 2 'I Don't Recall'
>
> 'I don't recall' is a classic cover for a probable lie and is used too often by dishonest people all over the world. In the US, Georgia representative Marjorie Taylor Greene used the phrase over 50 times in response to questions about her conduct and public statements in the run-up to the 6 January 2021 attack on the Capitol.[20] Nobody but Ms Greene can know for sure if she recalls or not, but if someone's memory of such a significant time is that poor, doesn't it tell us that they are unfit for their role, one way or another?

'White Lies'

The idea of a so-called 'white lie' – I put this in inverted commas because they don't really exist – is that the perpetrator believes it is harmless or in the interests of those who are lied to or the wider public. They don't exist except in extreme circumstances, such as telling Donald Trump the wrong nuclear codes. If a politician believes something is the right thing to do, they need to sell it for the right reasons. 'White lies' can come back to bite horribly.

> ### Example 1 Tony Blair's Lies About the Evidence of Weapons of Mass Destruction in Iraq in 2003
>
> A generous interpretation would say that the former UK Prime Minister thought it was in the public interest to lie. The Chilcot Report into the matter gave a damning conclusion that Blair deliberately blurred lines between what he believed and what he knew.[21] Arguably, this laid the foundations for the progressive collapse in political honesty that followed in the UK over the next two decades.

Subtle Twists

These are seemingly minor but highly significant alterations. It can seem pedantic to point them out, but they can create wildly false impressions.

As we've seen, falsehoods can happen by mistake, but one way of testing whether they are dishonest or not is by seeing the response when the errors are pointed out.

Example 1 Twisting the Facts

Former UK MP Claire Perry: 'The UK is getting 60% of its energy from renewables.'[22] Sounds great, and the only inaccuracy is that it was 60% of our *electricity* that was from renewables, whereas less than a quarter of our total energy supply was from low-carbon sources.[23] Since electricity supply is the easiest bit of our total energy to decarbonise, the reality is that while the statement makes it look as though we are almost on the home straight, in fact we're only just off the starting blocks when it comes to energy transition.

Failure to Correct Errors

Failure to correct the record distinguishes a lie from a possible mistake.

It is conceivable that the detail of whether the NHS figure was £350 million or £250 million could have been a mistake, but failure to loudly and clearly correct the record ensured that this was unambiguous deceit. The same can be said for any of the examples of deceit listed in this Appendix. Full Fact catalogues false statements made by MPs, invites them to correct the record, and reports on whether they have done so.[24] To pluck one example from this website, Dame Andrea Leadsom MP made several inaccurate claims during a television appearance in January 2024. Full Fact called her out on it and invited her to respond, but she did not.[25]

If you are wondering whether a UK MP has tried to pull the wool over your eyes or just made an honest mistake, there is a good chance you'll find your answer on the Full Fact website.

Staying Quiet

Bad men need nothing more to compass their ends, than that good men should look on and do nothing.

John Stuart Mill

This passive form of dishonesty is perhaps the weakest but most widespread type of deceit. There are some things that are not OK to stay quiet about, and ignoring your colleagues in Parliament lying to the public comes into that category. Similar cases might be an editor allowing popular but false journalism to stand, or someone keeping quiet about something that isn't right in the business they work for. Staying quiet undid the honesty credentials of every single sitting Conservative MP under the Johnson, Truss and Sunak governments, and every Republican who has served under Trump. It undoes any media channel that repeats known and provable falsehoods without challenge. I've heard it defended along the lines of 'Oh well, you have to, or it is the end of your career' – to which my answer is that it is obviously not OK to enable the demolition of democracy for the sake of your own career. Staying quiet is in clear breach of the last of the seven Nolan Principles of Standards in Public Life (Figure 38): leadership,

Example 1 Failure to Speak Out

Sadly, there are too many examples here to list individually. So many UK ministers have track records of everyday deceit which are so blatant that it is implausible to think that any MP in their party doesn't know they are being dishonest. So, every single one of them is guilty of staying quiet. MPs from several parties failed to call out the Brexit Bus claims, perhaps weakly justifying to themselves that a lie is OK if it gets the end result you want. It is not.

including challenging poor behaviour wherever it occurs. Loyalty to your party tribe is emphatically not an adequate defence.

Loopholing

This is honouring the letter but not the spirit.

Example 1 Hospitals

In 2020, Boris Johnson (then UK Prime Minister) committed to building 40 new hospitals by 2030. The public are allowed to have a reasonable image in their mind's eye of what a new hospital looks like. But in an 'Emperor's new clothes'-style twist on reality, the government classed refurbishments, new wards and new units in existing hospitals as 'new hospitals'.[26] This looks to me like flat-out Orwellian deceit.

Example 2 Tax Avoidance

Anyone who deliberately exploits a loophole to avoid paying tax ('tax avoidance' rather than the criminal 'tax evasion') is short-changing every member of society – including themselves in the long term. The UK Parliament website describes how 'aggressive or abusive avoidance, as opposed to simple tax planning, will seek to comply with the letter of the law, but to subvert its purpose'.[27] When we find ways to wriggle out of paying our taxes that we know subvert the intention of the law, we are directly contributing to poverty and lack of public services. We are undercutting the very society which has helped us to gain that wealth in the first place. The over-rich do not use public services, but their employees do, the people who maintain the roads they drive on do, the farmers who produce their food do. We need a culture change where both individuals and companies are proud to contribute to the society they live in through the payment of taxes.

Misdirection of Attention

This is a vital part of the toolkit of every magician. Focus people on the irrelevant stuff so they won't notice the important things that are going on.

Example 1 Air Travel
An airport loudly advertises the reduction in emissions from the airport itself and in doing so directs attention away from the dramatically more important emissions from the actual flights.

Example 2 Coal Mines
West Cumbria Mining's original environmental assessment of their proposed new coal mine discussed the emissions from extracting the coal, but not the much larger emissions from using it.

Example 3 Supermarkets
A supermarket proudly tells you about solar panels on its roofs, but not about the unsustainable foods on its shelves; or small reductions in plastic packaging, but not the carbon footprint of everything it sells.

Example 4 Oil
The oil company Shell produces a video of its chief executive talking about their renewable energy campaign with a reflection of a wind turbine in his eye, without mentioning that the expenditure on new oil and gas exploration far exceeds their renewables investment.[28]

Biased Gathering of Evidence

In this technique, a politician, a business or a journalist commissions research from people or an organisation that they know will tell them what they want to hear.

From time to time, I am approached by companies that want me to do some research in order to support what I know is a bogus narrative. We usually have a short conversation and that is the end of it. But they won't have too much trouble finding another consultancy that is happy to take the money and oblige. We've had a major tobacco company asking us to help them put together the case for sustainability of their vaping products; we've had companies hoping for support of their 'sustainable' beef burger chain; and we've been asked to research the carbon footprint of fighter jets. In principle, we are always available to offer supportive sustainability strategy advice to any company trying to do the right thing, whatever their industry. For Philip Morris International it would be how to shut down the company, for the burger chain it would be how to transition to a more plant-based diet, and for the defence company it would be advice on how to be sure that the business model was not incentivised to have a world in conflict, and what to do about what seemed to us to be highly credible claims of multi-billion-dollar corruption and multi-million-dollar funds for bribes, including payments for sex workers.[29]

Example 1 Mining

West Cumbria Mining paid for an 'expert' to rebut the challenge that Professor Rebecca Willis and I made to their plan to open a new coal mine. The person they chose had little expertise in either climate science or the steel industry for which the coal was supposedly needed. He had, however, spent much of his career promoting the Australian coal industry. By contrast, our subsequent counter-rebuttal cited evidence from internationally renowned economists, top steel industry experts and peer-reviewed climate science.[30]

> **Example 2 Food Industry**
>
> A study of over 1,400 articles published in the 10 most cited peer-reviewed journals on diet and nutrition in 2018 found that those which were funded by the food industry were 45 per cent more likely to conclude that a food product had health benefits, or to undermine evidence a product was harmful, than those studies which had not been industry-funded.[31]

Biased Selection of Evidence

This is the giving of preferential treatment to evidence from one side of the argument.

> **Example 1 Mining**
>
> To stick with the West Cumbria Coal Mine, when Michael Gove, as Secretary of State for Levelling Up, Housing and Communities, came to review the case for the mine, he gave preference to the evidence from non-expert reports written by those under the pay of West Cumbria Mining over what was by then a wealth of evidence from highly respected, credible and pro-bono sources covering the steel industry, climate science and economics. Why did Gove privilege the bogus evidence over the good stuff? Was it incompetence or dishonesty? Many people say what a clever man he is. I can't prove dishonesty over incompetence, but I can let you make up your own mind. And either way, to me it surely demonstrates unfitness for the job.

> **Example 2 Restricted Testimony**
>
> Climate protesters in the UK have been instructed not to tell juries the reason behind their actions and not to mention the climate emergency when defending themselves.[32] This is akin to someone

breaking a car window on a hot day to rescue a baby at high risk of dying inside the car, but not being allowed to say in their defence why they broke the window. It is for the jury to decide whether the motivations behind an action constitute lawful excuse, not for the judge to prevent the jury from hearing the motivations at all. Furthermore, jury members have a right in law to acquit a defendant according to their conscience.

Altering History

Winston Churchill famously said, 'History will be kind to me for I intend to write it.' Orwell also addressed this form of deception in his novel *1984*, in which the 'Ministry of Truth' was dedicated to rewriting or creating articles and evidence to suit its present position. Sadly, fiction has become fact.

Example 1 Retrofitting

Dominic Cummings said in 2020, 'Last year I wrote about the possible threat of coronaviruses and the urgent need for planning.' In fact, he retrofitted his own blog to make it look as if this was the case.[33]

False Impression

This is giving the appearance that someone or something is genuine, when they are (or it is) not.

Example 1 @factcheckUK

In November 2019, Boris Johnson faced Jeremy Corbyn in a televised election debate. Just before the event, the Conservative Party changed the name of its Twitter account from @CCHQPress to @factcheckUK, with a clear intent to disguise party messages as an impartial fact-checking service. Twitter called them out for it, and it hit the headlines. Giving generous benefit of the doubt, all this might have been written off as a mistake from an over-enthusiastic member of the party headquarters team. However, according to my honesty assessment criteria, James Cleverly provided all the evidence we need to know that *he* is unfit to be in Parliament when he refused to acknowledge that this was a problem when challenged directly on Newsnight.[34]

Example 2 Brexit and the NHS

We have already mentioned the Brexit battle bus, with its slogan:

We send the EU £350 million a week – let's fund our NHS instead.

Much more serious than the £350 million only actually being £250 million, there was clearly a false impression here that leaving the EU would generate a net saving that would be available for the NHS. Instead, the reality was always going to be a high cost associated with leaving. In the event, we left the EU, and the NHS has been drastically underfunded. Not only was there a lie in the numbers, but there was also a false impression that the intention was to divert any savings to the NHS.

Example 3 Oil Drilling

The Willow Project is a planned oil drilling project by ConocoPhillips in northern Alaska. Among the many false impressions is the trivialisation of planned emissions, with the true statement that it 'will create only a fraction of 1% of all U.S. emissions' – as if that isn't a huge amount. (They also say that if they don't produce the oil, US citizens will have to import it – making the false presupposition that oil consumption must continue at today's rate. There is no acknowledgement that energy demand reduction is a necessary part of any sustainable future, nor that renewables could meet much of that demand.[35])

Fake Judgement

This is saying you believe something that you do not actually think is true.

Example 1 Hiding the Truth

Back to the Cumbria Coal Mine, in 2022 Robert Jenrick (Secretary of State for Housing, Communities and Local Government at the time) refused to call in the planning permission on the mine, claiming it was an issue of only local significance.[36] Yet he knew the world was looking on in the run-up to the UK hosting COP26, and also that the millions of tonnes of emissions at stake were globally significant. So that is flat-out dishonest.

Hidden Motive (or Influence)

This can be either acting solely for the purposes of a hidden motive or influence, or even just not coming clean about reasons and connections which would be likely to have an impact on your views or behaviour.

Example 1 On the Take

Former Conservative MP Scott Benton was caught in a newspaper sting,[37] offering to table parliamentary questions and leak confidential policy documents to the gambling industry in return for payments of £4,000 per month.

Example 2 Revolving Doors ...

Former Chancellor Kwasi Kwarteng used to be a paid consultant at Odey Asset Management. When he launched his disastrous mini budget, his former employer made money betting against the British pound. I'm not saying Kwarteng launched his mini budget to make his old boss some money, but the connection stinks.

Burying Bad News

Any time a 'royal scandal' is out, I wonder what news it is really burying. This category overlaps with misdirection of attention, but it deserves its own mention. 'Bad news' is often released just before a school holiday in the hope that people will be too busy to pay attention.

Example 1 LTNs

In 2023, Rishi Sunak ordered a review of Low Traffic Neighbourhoods (LTNs) as part of his narrative of being 'on the side of drivers'. In 2024, the government tried to bury this report, as its finding was that most people supported LTNs.[38]

Example 2 PPE

£1 billion worth of unusable COVID-19 PPE was burned by March 2023.[39] The UK government quietly slipped this out just before the Easter holidays.

Camouflage

This is the tactic of including a significant, highly impactful change behind a cover of plausible actions.

Example 1 Public Order Bill

The 2023 UK Public Order Bill was described as 'deeply troubling' by the UN Human Rights Chief, who called for it to be reversed.[40] Rishi Sunak's government described the Bill as protecting 'hard-working' people and allowing the police to take a 'proactive' approach:[41] making it legal for police to stop and search individuals without any suspicion and curtail both the movement and

> internet rights of someone who has never been convicted of any criminal offence. It also allows the police to arrest people *before* they have joined a protest, on suspicion that they will cause disruption. Curtailing the basic human rights of citizens is no good to anyone but those hanging on to a power they have not earned through thoughtful and responsible public service.

I could expand this taxonomy into an encyclopaedia, but I hope by now I've made the point that deceit is a lot more than just lies. The perpetrators of all the examples here are unfit for political life. If you are reading this but don't live in the UK, I'm sure you will have many of your own examples from your own country. The point is – always be on the lookout for 'untruthiness', wherever you are.

APPENDIX 2   HONESTY AND TRUST CRITERIA

Five Criteria for Assessing a Politician's Honesty – in More Depth

Here, in more detail, are the criteria described in Chapter 7.

For each one, a politician can have a score of 'Terrible', 'So-So' or 'Exemplary', and clearly there are shades in between.

Here's a guide to the significance of those scores.

- **Terrible:** If anyone scores Terrible on any one of these criteria, it is a show-stopper. They are probably unfit for office, even if exemplary on every other count. Only vote for them if the alternative is even worse.
- **So-So:** This will just about do in the absence of anyone better.
- **Exemplary:** If all politicians were like this, people and planet could thrive.

For each criterion, I have described what Terrible, So-So and Exemplary look like, and I've given one or two real-life examples when I think illustration is helpful. Actually, I've found it difficult to pick examples from such a depressing plethora of deceit. Again, my examples are mainly from the UK – but I'm sure you can think of more local examples if you live elsewhere.

Here are the criteria.

(1) Honesty Track Record

To what extent has this person been consistently careful to impart the best attainable view of reality as they see it? Or have there been one or more indisputable examples of them saying something in relation to a policy issue that they know to be incorrect? Have they said things that they know are misleading? Have they deliberately used any of the techniques listed in the Taxonomy of Deceit (Appendix 1)?

If they have been found to have made an error of fact, have they taken care to correct the record, not just quietly in the small print, but clearly, and seeking, as far as is practical, to ensure that all those they have imparted false impressions to have been put straight? FullFact.org catalogues clear-cut untruths from MPs, informs them, and then logs the extent to which they subsequently correct the record. It's a fantastic source, although it only covers clear lies rather than the full taxonomy of deceit. In 2023, it called out 34 MPs for failing to correct the record. Diane Abbott, Andrew Bridgen, Stephen Barclay, Chris Philp, Keir Starmer and Suella Braverman all failed to correct their errors more than once – top of the list was former Prime Minister Rishi Sunak, with five failures to set the record straight.[1]

Terrible: Has deliberately misled the public on at least one significant occasion for political gain, either by lying or deploying another tool of deceit, and has failed to correct the record when challenged.

Examples: Grant Shapps has such a long history of dishonesty that it is difficult to choose which examples to cite. As well as those I have already mentioned in Appendix 1, Shapps also wrongly stated that the government's proposed legislation on curbing the rights of workers to strike would bring the UK 'into line with' other European countries.[2] Furthermore, back in 2005 he was using the pseudonym 'Michael Green' in his job as a multi-million-dollar web-marketer, at the same time that he was an MP. He denied this for years, until pictures emerged to prove it.[3] As I mentioned elsewhere, in *The Rise of Political Lying*, Peter Oborne writes that Tony Blair told many lies, starting early in his career, including lying on his CV before becoming an MP – a forewarning, perhaps, of honesty problems over weapons of mass destruction and the disastrous consequences that followed.[4] Boris Johnson has told so many lies there is an entire website dedicated to them.[5]

So-So: Can be careless over facts. Corrects the record but does not take enough care to ensure those they have misinformed have been put straight. Sometimes disingenuous about motives and reasons behind decisions or opinions. Sometimes guilty of a degree of spin and misdirection of attention but never clear-cut untruthfulness.

Exemplary: Gives us the clearest view that they can of issues as they genuinely see them. Is honest about the reasons for their opinions and political positions. Corrects the record clearly when mistakes are pointed out, with apologies where appropriate.

Example: Everybody makes mistakes; the key thing is to own up to them. Labour MP Chris Bryant wrongly claimed that there was 0 per cent VAT on heat pumps in the EU. He issued a public correction on X, thanking Full Fact for bringing this error to his attention.[6]

(2) Transparency of Funding, Interests and Influences

Are their sources of funding, both personally and politically, made transparent? MPs have to declare any sources of income to the Register of Members' Financial Interests. This is supposed to include everything that might 'reasonably be thought to influence their decisions as an MP'.[7] But this isn't necessarily the whole story, as money can be supplied through indirect means. *Transparency* of association is all about clarity over who they meet with, and whose advice they seek, as well as what is said at those meetings. This too can be murky, as the boundary between social events and business blurs. I don't think we should be putting our noses into politicians' private lives – unless there is actually a work element taking place, in which case it absolutely does become our business. If a minister attends a media baron's wedding, for example, there is a good chance that there is a business element to it.[8] Tortoise Media (https://tinyurl.com/wm-accounts) has a website documenting how MPs are funded.

A short time browsing through it will give insights into corrupting influences.

Terrible: Has major direct or indirect sources of income that are not clear; and/or has covert discussions with interest groups either formally or informally.

Examples: As I mentioned in Appendix 1, in December 2023, in a *Times* newspaper sting, the Conservative MP Scott Benton offered to lobby ministers on behalf of the gambling industry for a fee of up to £4,000 a month.[9] Michael Gove, as Education Secretary, was receiving payments from the Murdoch-owned *Times* for a newspaper column, and monthly payments from the Murdoch-owned Harper Collins publishers for a book that he never wrote, while also attending Rupert Murdoch's private functions and simultaneously pushing an agenda for digitising education, from which Murdoch planned to derive a significant proportion of his empire's profits.[10]

So-So: Has sought to downplay some sources of funding or influence, whether direct or indirect, but does not actively conceal.

Exemplary: Open about all their sources of income both personally and politically. Open about who lobbies them and whose advice and evidence they seek.

Examples are hard to find, because how would we know?

(3) Freedom from Corrosive Interests

Being transparent is an essential step, but it isn't enough. We also need to know that the funding and lobbying they receive doesn't stand to influence their political decisions against the public interest. And we need to know that the people they meet and those who advise them also have high standards of honesty. Only by asking these questions down at the next level of the funding and influence supply line can we start to unpick some of the more sophisticated and indirect ways in which malevolent parties sometimes seek to shape the agenda.

Terrible: Regardless of transparency, this person receives significant funding either personally or politically from sources that have vested interests; and/or they allow undue influence of interest groups in return for benefit in kind other than the wider public interest.

Example: Andrew Rosindell MP was among a number of British MPs who regularly accepted expenses-paid trips to the Cayman Islands, a notorious tax haven helping multi-national corporations to underpay corporate income tax. The Labour Peer Margaret Hodge has said £290 billion per year is lost to the UK economy through economic crime. Following a trip to the Cayman Islands in 2018, Andrew Rosindell pushed back against increased transparency measures through votes, comments, letters and social media.[11]

So-So: Allows themself to receive some funding and other favours from sources that are motivated by the benefit they might secure either implicitly or explicitly in return. Indulges in wishful thinking about robust barriers between these funds, favours and their own political behaviours.

Exemplary: Ensures that their funding comes without strings – either formal or unspoken. Careful not to accept money and not to curry favour from sources that may entice them to act other than in the public interest.

(4) Scrutiny of Information Sources

Do they select the highest-quality and most robust evidence, or do they pick their sources to give the result they want to see? Do they favour impartial expert evidence that is not in the pocket of financial or political vested interests? When they commission research or advice, how do they select their researchers and advisers? Are they looking for truth or for something that will provide cover for the action they have already decided on? Do they even have the critical thinking awareness to understand the biases that may lie behind the information sources they receive?

Terrible: Selects sources of evidence and advice that suit their pre-existing agenda. Gives priority to evidence that suits their preferred position over more robust and impartial evidence from elsewhere. And/or does not demonstrate the ability to discern credible, reliable and trustworthy sources from those that are inexpert or motivated by particular private or corporate interests.

Examples: Michael Gove (again) and the West Cumbria Coal Mine. Gove ignored government business department papers that said the UK steel industry was likely to decarbonise by 2035 and instead used steelmaking as an argument for the mine.[12] Rishi Sunak commissioned a report on Low Traffic Networks but buried it when the finding turned out to be that LTNs were both popular and effective, contrary to the government's pro-car policies.[13]

So-So: Can be careless over sources. Is not sufficiently aware of the need to take account of the motivations, expertise, track record and political inclinations of sources when assessing their evidence. Does not, however, deliberately seek biased evidence.

Exemplary: Critically reviews all their sources. Asks carefully, 'How do I know I can trust this?' Seeks to understand and take account of the motivations, level of expertise, political inclinations and honesty track record of sources. Seeks impartiality, even if the result may be inconvenient or uncomfortable.

(5) Expecting Honesty from Others

This fifth criterion is about whether they honour their responsibility to help uphold standards in public life. Are they happy to stand by while others are dishonest, or do they make a stand for honesty even if it is unpopular, or even if it means calling out someone in their own party?

When something untrue is said in support of their agenda or their party, do they politely but clearly and publicly point out the error? Or do they stay quiet while others are being dishonest in the name of a cause or party that they support? At worst, do they bask in the perceived benefit to their cause of others deploying the taxonomy of deceit to bolster support for their agenda?

By having the expectation that our politicians uphold high standards in others, we create an environment in which our political figures know that if they are careless with the truth they will become isolated. They will become a reputational risk to all those around them.

Terrible: Supports and repeats probably bogus narratives. Fails to call out colleagues who have misled on significant issues. Votes in favour of colleagues who they know have deliberately and repeatedly misled the public on matters of political significance.

Examples: Any Conservative MP who voted for Boris Johnson to become party leader after his '£350 million a week for the NHS' slogan had been exposed as bogus, yet Johnson continued to push it. Or any MP who endorsed any of the proven and uncorrected mistruths documented by Full Fact or the boris-johnson-lies website. Anyone who failed to call out in a timely manner Trump's evidence-free claims that the 2020 US election had been rigged.

So-So: Does not repeat or endorse statements that they know to be false or misleading. Will not serve under a probably and serially dishonest party leader without publicly calling this out. Has sometimes stayed quiet in the knowledge that colleagues have been careless with the truth.

Example: Conservative MPs who tolerated Boris Johnson as an MP but resigned from his cabinet when he became party leader.

Exemplary: Requires honesty and integrity of their colleagues in all parties. Is prepared to call out and stand against misinformation even when it comes from within their own party and even when it is politically inconvenient to do so.

Examples: Caroline Lucas MP has always been very clear about how important honesty is to her, even writing a letter to the Speaker of the House, signed by other MPs from different parties, condemning then-Prime Minister Boris Johnson's repeated misinformation to the House of Commons.[14] As I noted in Appendix 1, Dawn Butler MP called out Boris Johnson for lying and was ordered to leave the chamber of the House of Commons for doing so.[15]

Five Trust Tests for a Person or a Decision

Competence wasn't included in the five honesty criteria, but when it comes to evaluating a specific decision, competence has to be part of the equation. We may also want to ask what others (who we trust) think of the trustworthiness of the decision-maker. It can be useful to work out for ourselves whether the logic of the decision makes sense, but it is super-important to be aware of our own limitations here, as lay people, on decisions that require detailed knowledge and expertise. Remember how easy it is for a dishonest politician to make a bogus argument look plausible to even the smartest member of the public.

So here is a different model for making up our minds on a specific decision. It is important to note that these five trust tests emphatically do *not* include charisma, charm or wit. These can easily be used to pull the wool over our eyes.

One of the worst judgements I can personally think of making over a political decision was to back Tony Blair over the Iraq War. Going over these five criteria helps me understand how it happened. He was new in office and as far as I could tell had no track record of deceit. Others similarly rated his integrity at the time. I couldn't see a vested interest that might divert him from the national or international interest. He seemed very competent. In terms of whether it made sense to me, the decision to go to war with Iraq looked crazy, but I thought, 'OK, you have access to more information than me. You seem smart, well-intentioned and better placed than me to decide. I'll trust you, even though it goes against my instincts.' If I'm being charitable, Blair made the catastrophic decision that a white lie would be in the world's best interests, and exaggerated the evidence for weapons of mass destruction. I never trusted him in the same way again.

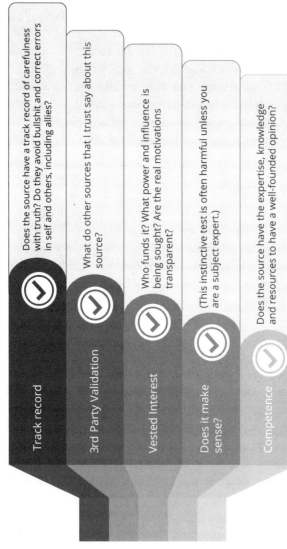

Figure A2.1 Here is an example of the tests in action.

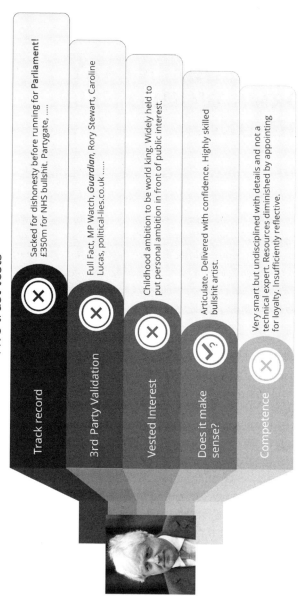

Figure A2.2 How Boris Johnson measures up on the trust tests: my personal view. Image credit: Isabel Infantes/AFP/Getty Images.

APPENDIX 3 WHAT DEMOCRATIC AND PARLIAMENTARY REFORMS WOULD HELP IN THE UK?

You don't have to be an expert on democratic and parliamentary procedure to see that the UK system isn't currently set up to deliver the high-quality decision-making that we need. In fact, things are so broken that the fresh eyes of a relative outsider like me might even be helpful for seeing what change is needed. I have kept things high-level, drawing on ideas from people who are closer to the workings of the UK's politics than I am.[1]

In a thriving democracy, transparency and honesty are maintained by healthy institutions, by a free and responsible press, and by an engaged, informed and participating public. But just when we need to be raising the effectiveness to a level never seen before, in the UK (and elsewhere), all these elements have collapsed to a worse state than I thought I'd ever see. The right to non-violent protest is being hacked down before our eyes. Meanwhile, key institutions are having their independence undermined by the systematic and increasingly cynical insertion of partisan people into top roles. We don't have an adequate process for selecting the best candidates to become MPs, and we don't make it possible for most of the other best potential candidates to join the running.[2] In Parliament, the decision-making processes are often desperately poor, only in part because of the inadequate public involvement. If you are unsure about any of this, then a few memoirs by former MPs will leave you in no doubt.[3]

Long-standing problems with the voting system mean that even at election time, most people have little meaningful say.

Electoral Improvements

- **Replace the first-past-the-post** voting system with either proportional representation or, much more simply, a single transferable vote in each constituency. The current system has logical flaws that are practically impossible to defend except with bullshit. To win, you need to have nobody else splitting your vote. A politician gets elected not by being the one who best represents the will of the people, but through carefully adjusting their policies and espoused values so as to be the least-bad option in the eyes of the largest proportion of the voters. The candidates have little incentive to appeal either to those for whom they will be the least-bad option, whatever they do, or to those for whom they will never be the top choice, however hard they try. The most successful strategies are those that focus on the swing voters in marginal seats. This is just a small proportion of the electorate, leaving the rest of us without influence. The first-past-the-post system also means that if two or more candidates stand for something that the public overwhelmingly wants, they are all perversely likely to lose out to a candidate with a minority view but no competition. Quite clearly this does not represent the will of the people anything like as well as a single transferable vote system, in which you vote for whoever you most want and then rate the other candidates in order of preference. If your preferred candidate gets knocked out, your vote transfers to your second and then third preference. This means you can vote for your dreams without wasting your vote. It means that even if you can't have your ideal candidate, you can still have just as much influence as anyone else over the choice between the final two. It means there can be healthy competition between candidates in the same political space. In the UK, it would be incredibly simple to implement. There would be no need for any reworking of constituency boundaries. The votes would, admittedly, take a few hours longer to count. The result would be a radical improvement in the extent to which politicians represented the will of the people at both the national and the local level.

- **Tighten the cap on electoral campaign spending** to level the playing field. In the UK, only one party has ever come close to reaching the limit that was set at £19 million. But in 2024, that same party almost doubled the cap to £36 million, knowing that they are the only party likely to benefit from the increase, having secured £19 million from just two donors, both of whom, as it happens, were awarded knighthoods by Rishi Sunak.[4] This pales into insignificance compared with the amount of money spent on elections – at all levels – in the US, to say nothing of the amount of money spent by lobbyists in Washington.

MPs' Conduct and What They Should Be Able to Expect in Return

- **Introduce a parliamentary code for all MPs** that is just as exacting for backbenchers as the ministerial code is for ministers. The code should not only call for MPs to be honest but extend, along the lines of the Nolan Principles, to include notions of transparency, openness, freedom from corrupting interests and upholding standards in others. The code should be enforced not by the Prime Minister, who as we have seen may not have impeccable standards of their own, but by an independent body which perhaps includes randomly selected lay people to help judge MPs' conduct.
- **Tame the whipping system.** This is the system by which MPs are told by their party how to vote at almost every turn in Parliament, and what to say in public. The whips are currently responsible for both coordinating voting and MP welfare, so that they have enormous leverage over the life of an MP. For anyone

in doubt as to the corrosive influence of the whips, as they currently operate, I recommend Rory Stewart's account of his time as an MP, detailed in his book *Politics on the Edge*.[5] There is value in the whips encouraging coordinated activity from government, but there are two lines that should never be crossed. The first is that no MP should ever be asked to vote against their conscience. The second is that they should never be asked to say something that they do not believe; they should, of course, never be asked to be dishonest. In other words, there is a place for an MP to vote against their own best judgement for the sake of unity and coordination of government. (This is different from doing so out of party tribalism, which is fundamentally unhealthy.) But there is no place for being asked to say that you believe it is the right thing to do when you don't. Pressuring someone to do so should be a breach of parliamentary standards, just as many other forms of coercion are illegal in public life. Also, the roles of vote coordination and welfare should be separated, probably along with improvements to the welfare role.

- **Tighten the rules on lobbying.** Make it, for example, an offence either to lobby or to be lobbied without transparency, or to remain quiet about knowledge of corporate lobbying that is not transparent. We need restrictions on lobbying by former MPs within a certain time period of leaving office. We also need restrictions on former ministers taking up roles in companies for which they had governance responsibilities while in office, so that the anticipation of being offered such a role in the future does not corrupt ministerial behaviour.
- **Reform of both pay and expectations of MPs.** Doing the job of an MP properly is hard work and very full-time. Realistically, it is a lot more than 37.5 hours a week. When you work that hard, you need everything else in your life to be a bit simpler, and that is easier if you have a bit more money. I think it is reasonable for MPs to be able to afford cleaners, gardeners and someone to iron their clothes if they like, without having to think too hard about it.

Without being reckless, it is reasonable for them to be able to afford to easily make the most of their time off with friends and loved ones. The salary should also reflect how hard they have to fight for their jobs every four or five years, and the intrusions, annoyances and even dangers that come with it. So, we should probably be paying them a bit more than they currently receive (£91,346 in June 2024). In return, we should expect that they *do* work full-time, do not have significant other jobs, and do not receive payments for anything that could corrupt their motivations. Their staff should similarly be decently paid and on secure contracts. This will help free MPs from over-reliance on outside advice, which can so easily lead them into the pockets of industry lobbyists.

- **Respect from the public and the media.** Although it is legitimate to forcefully insist on integrity from our politicians, even this can be done respectfully, staying out of their private lives as long as they remain legal, leaving their families alone, talking and writing about them with respect even when we are furious with their views, their policies or their behaviours. While we do need to see competence, we need to understand that the issues are complex, and the arguments may be difficult to explain. We need to allow MPs to change their minds, either in the light of new evidence or simply because they wake up at three in the morning realising something new. We need to understand that they are bound to make some mistakes, and when they do, if we want them to own up, we had better show some understanding.

Treating MPs with respect may be the hardest of my suggestions to put in place, since it requires the cooperation of the whole of the public and the media, including the UK's famously vicious tabloid newspapers. Right now, it is an especially big ask of the UK public, since we have so much cause to be furious, having witnessed such an epidemic of appalling behaviours from so many MPs over the past decade or more: 'Partygate', routine deceit (of which the Brexit Bus is just one icon), deals for mates, corrupt lobbying and so on. But if we want to see things get better, the way to respond, hard though it can be in practice, is, as Michelle Obama famously put it, 'When they go low, we go high.' In other words, it is not a contract in the sense of we will respect you as long as you respect us. It is more like, 'We will respect you whatever you do, but if you don't reciprocate, we will find someone else to represent us.'

Every one of us can help put this culture in place through the way we talk about politicians and through the media we support. I admit I have not always managed to honour this principle in private.

Reform of Appointments

Reform of public appointments is necessary to improve the independence of the selection process, including, most notably, nominations for peerages to the House of Lords (leaving aside the question of complete reform of the second house!), and key media and regulatory roles such as in the BBC, ITV, Channel 4 and Ofcom. It is technically already illegal to promote someone to the House of Lords, or give a knighthood or damehood, just because they have given money to your political party. In practice, however, this can be done as long as there is no formal written contract that the donation is in exchange for an appointment.

The House of Lords does appoint a few cross-bench members, but I even found out first-hand how corrupt this procedure can be when I was offered a service to have my own chances improved should I wish to apply.[6] By paying £20,000, I could supposedly have nice references organised in exchange for an understanding that I would do some mutual back-scratching once appointed. This service was justified to me as being no worse than paying for coaching to help your child to get into a good university. The basis for an honour should be services to the country, and the basis for membership of the House of Lords should be the contribution you can make to the world once you are there, not a bought position for individual gain.

Tools for Better Decision-Making

- **Create a unit for joined-up government.** The role of this unit would be to map policies across departments to see overlaps and conflicts. Tony Blair had a Strategy Unit along these lines, whose job it was to give knotty issues to an interdepartmental team. The Cabinet Office currently fulfils this role to some extent in principle, but it is dependent on the Prime Minister's governing style, and as we have seen, that cannot be relied upon, especially when that person has been selected by the small and very unrepresentative group of the party membership. In all my work, with all kinds of organisations, the challenge of joining things up across departments is always hugely important and difficult. Most of them seem better able to do this than the UK government, for whom the need is even greater yet the processes hopeless. While a minister needs detailed understanding of their own area (which is partly enabled by keeping them in post for long enough to get their heads around it), they should also be expected to care equally about the work of other departments. The job of this unit would be to point out early on where the dots aren't yet being joined.
- **OnePlanet** is an interesting example of a digital tool to help with this.[7] It allows different organisations and departments to enter their plans and strategies, and then to see, visually, how they join up with each other, or what needs to happen to make them do so.[8] It is being used by local and regional government, as well as community groups and companies. Founder Pooran Desai is now also starting to work with central government. I can't tell for sure how well this will work at national government level, but it looks promising and, certainly, it is helping some of my clients to join the dots of complex multi-stakeholder strategy. Pooran Desai, who has an impressive track record as a sustainability pioneer, puts it like this:

Everything is interconnected. We delude ourselves if we think we can solve the climate, economic and health problems we face in silos. We need joined-up solutions and OnePlanet enables users to collaborate intuitively to find the solutions. It actually makes life a lot simpler. It uses a different part of our brain – the right hemisphere which can see and understand the big picture and which is underdeveloped in modern Western society, leading us to a literally delusional relationship with reality.

Sustainability and creating a regenerative future is all about understanding this very simple idea – our health is dependent on climate, which is dependent on the economy and so on. Yet we don't have the tools to visualise and manage for interconnectedness. My personal mission is to rescue people trapped in Excel spreadsheets so they can start seeing the bigger picture and reality as it is. What's not in your Excel spreadsheet is as important as what is – in fact it's probably more important.

OnePlanet uses so-called graph, or network, database technology which enables our users to organise information on the basis of its interconnectedness. Users can convert any project, plan, strategy or policy into a common 'atomic structure', or building blocks, of Outcomes (what you want to achieve), Actions (what you will do to achieve them) and Indicators (what you might measure and track). Once in this common structure, users can see how internally joined-up they are, but also enable any number of projects, plans, strategies and policies to be interconnected so users can easily identify synergies, opportunities, gaps and conflicts, collaborating in ecosystems across traditional silos of different sectors, organisations and departments.[9]

- Increase the public role in thoughtful, joined-up decision-making. This can be done, for example, through Citizens' Assemblies and Citizens' Juries (see Box in Chapter 7).

These have been among the most fruitful and heartening recent innovations in our democratic process. In the UK, the missing link so far is that they don't have enough clout. The missing piece in the puzzle is to give Parliament a duty to act on their findings – unless, for example, it can demonstrate a more careful, more robust and equally transparent process that suggests a different course of action.

APPENDIX 4   RESOURCES

The lists here are obviously not exhaustive and come with apologies to all the fantastic media, fact-checkers, books, films and other organisations that I have missed out. But everything I've included here passes the credibility criteria that I've outlined earlier to some extent, and most of them with flying colours, as far as I can tell. And I've written a bit about why I think they are interesting and sometimes why caution is also needed. Please use it, of course, to create your own better lists.

News

The BBC and their BBC Verify page: www.bbc.co.uk/news/reality_check

Standards have slipped badly at the BBC, following sustained attack (see Chapter 8). Still a valuable source of information, but now must be viewed with eyes wide open to what is *not* being covered and how the flavour is influenced by the many Conservative Party allies in top roles. Remember also that components of the Polycrisis are generally covered in isolation, and treated as a side issue, if at all.

Channel 4 News, including their FactCheck page: www.channel4.com/news/factcheck

Of the main UK news channels, Channel 4 stands out for its hard-hitting journalism. Channel 4 was established as a public service broadcaster, but, unlike the BBC, it receives no public funding and instead relies on commercial activities, including advertising. Channel 4 News, its flagship news programme, is known for its rigorous interviewing style and commitment to balanced reporting. As I write this, they have been covering climate and other environmental issues with more rigour and seriousness than the BBC. In 2005,

Channel 4 News set up FactCheck, a platform dedicated to verifying the accuracy of claims made in public discourse. It focuses on analysing statements by politicians, celebrities, and other public figures and sources. FactCheck employs a rigorous methodology to assess the veracity of claims, drawing on official data, academic research and expert opinions. Their findings are presented transparently, with clear explanations of the evidence used to support their conclusions.

The *Guardian*: www.theguardian.com/uk

The *Guardian*'s reporting style prioritises factual accuracy and in-depth analysis, often presenting a centre-left perspective. Owned by the Scott Trust Limited, a unique non-profit structure, the *Guardian* is relatively free from the influence of shareholders or private owners. Instead, profits are reinvested into journalism, ensuring editorial independence. The climate emergency is given a more serious treatment in the *Guardian* than in any of the other broadsheets. It dedicates significant resources to covering the issue, offering in-depth analysis, investigative pieces, and dedicated sections on their website and print editions. This commitment reflects its broader mission of informing the public and promoting social justice. It is particularly known for its invaluable investigative journalism, including the Panama papers, the phone hacking scandal and the Cambridge Analytica/Facebook data breach.

Financial Times: www.ft.com/

A for-profit publication owned by the Japanese company Nikkei, which is owned, in turn, by its employees. Both the *FT* and Nikkei stake their entire reputation on the trustworthiness of their content, for which they are both highly regarded around the world. The *FT* Editorial Code requires all contributors to report any errors they subsequently find in their work, to respond to queries that arise, to take great care with their sourcing and attribution, to declare interests and more besides. To my mind, it is trapped in outdated traditional economic thinking that we must learn to think beyond, as it can't deal with the Anthropocene challenge. But with that caveat, and provided, as you read, you are able to pull out and correct for the presuppositions about how the world has to operate, it is a very reliable source.

Byline Times: https://bylinetimes.com/

Set up by journalists Peter Jukes and Stephen Colegrave, *Byline Times* aims to deliver investigative journalism that goes beyond what is covered in the everyday news cycle. It has attracted top-class journalists, including Adrian Goldberg, John Sweeney and Peter Oborne, the last of whom spent years as a senior reporter in the conversative media. I have found it an important source of news that you might not find elsewhere. It's an offshoot of Byline, a crowdfunding platform for journalism projects. This means that *Byline Times* relies on reader contributions to fund its in-depth reporting, often focusing on under-reported stories and holding powerful figures accountable. *Byline Times* is not politically partisan; it values neutral, transparent and fact-based reporting, and makes a clear distinction between fact and opinion pieces. The climate emergency is a core topic, with a prominent section featured on the website.

CarbonBrief: www.carbonbrief.org/

A valuable source of news related to climate science and policy, including the latest research, policy developments, international climate negotiations, and the impacts of climate change on society and the environment. CarbonBrief aims to provide clear, accurate and timely information to help inform the public debate on climate change. It is known for its in-depth investigations, fact-checking, and explainer articles that break down complex topics into accessible content. The organisation is funded primarily by the European Climate Foundation and through reader support. I do not always agree with their *interpretation* of data, especially in the light of the global dynamics of energy and climate as outlined in this book.

Declassified: www.declassifieduk.org/

Established in 2019 by journalists Matt Kennard and Mark Curtis, Declassified UK is an independent investigative journalism organisation that focuses on the UK's role in global affairs. Its reporting delves into the activities of British foreign policy, military and intelligence agencies, and their impact on human rights and the environment. It is funded through a combination of public donations, trusts and foundations, which are listed on its website. It does not accept funding from governments or corporations.

NPR – National Public Radio: www.npr.org/

US-based, non-profit membership media network, providing news, cultural programming and entertainment to over 1,000 public radio stations across the US. Throughout its history, NPR has garnered a reputation for excellence in journalism, earning numerous awards for its investigative reporting and innovative storytelling. Perhaps a measure of its balance is that NPR has faced allegations of political bias from all sides, and specifically, criticisms from both sides regarding its coverage of the Israel–Palestine conflict. NPR broadcasts and publishes comprehensive reporting on a wide array of topics, including national and international news, politics, business, science, arts and culture. Besides its radio broadcasts, it also produces a lot of content for its website, podcasts and other media channels. A significant portion of its revenue is derived from programme fees and dues paid by member stations, which, in turn, receive funding from listener contributions, corporate sponsorships and grants.

Democracy Now! www.democracynow.org/

US-based *Democracy Now!* started as a radio broadcast in 1996 and has since expanded into an hour-long TV, radio and internet news programme airing each weekday. In addition, its multi-media presence now includes additional web content and podcasts. *Democracy Now!*'s content is focused on current affairs and covers a wide range of topics, including politics, human rights, environmental issues and grassroots activism. It also aims to function as a watchdog regarding the effects of American foreign policy. Key funding sources include foundations that support independent media and social justice initiatives, such as the Park Foundation and the Lannan Foundation. *Democracy Now!* maintains a strict policy of not accepting funding from corporations, governments or advertisers, to ensure its editorial independence.

Al Jazeera: www.aljazeera.com/

Established in 1996 by the Qatari government, Al Jazeera has become a prominent global news network known for its in-depth reporting, particularly in the Middle East. They have provided a platform for voices often unheard in Western media, challenging traditional narratives and offering alternative perspectives on current affairs. However, Al Jazeera's state funding raises concerns about editorial independence and potential bias reflecting Qatari foreign policy interests. Al Jazeera's reporting style has also been described as occasionally sensationalised. Despite these criticisms, the network remains a significant force in global news, offering a unique perspective and shaping the conversation on international affairs.

New York Times: www.nytimes.com/

Known for its in-depth reporting, insightful editorials and impactful investigative journalism, the *NYT* is owned by a publicly traded company with a dual-class share structure. This grants control of the majority of the board of directors to the Ochs Sulzberger family, who have owned the paper for over a century. Its reputation for getting its facts right and its impartial journalism has been tarnished, however, for example by its failure to retract stories of mass rape in Israel that turned out to be false, and by firing the hugely respected journalist Chris Hedges after his refusal to go along with its pro-Iraq War stance.

Double Down News: www.doubledown.news/

An independent online video outlet that prioritises editorial independence by relying solely on reader contributions through Patreon, a crowdfunding platform. This allows it to avoid advertising revenue, ensuring its content remains free from potential commercial influences. Contributors to DDN have included prominent figures from the radically anti-capitalist George Monbiot to the instinctively conservative Peter Oborne. The outlet's content leans towards a leftwing political perspective and appears to be a bit more overstating in the language and imagery used than its more neutral or centrist counterparts. To me, it would be more powerful if sources were referenced. Founded in 2017 by Yannis Mendez.

The Real News Network: https://therealnews.com/

Founded in 2007 by documentary filmmaker Paul Jay and journalist Mishuk Munier, The Real News Network (TRNN) is a non-profit news organisation. Its coverage spans national and international affairs, often featuring five-to-seven-minute investigative reports. TRNN was started with the mission to unpack complex issues and present them in a clear and accessible way for a broad audience. Funding is mostly derived from viewer support and grants. Pulitzer Prize-winning journalist Chris Hedges (see *New York Times* above) contributes extensively.

OpenDemocracy: www.opendemocracy.net/en/

OpenDemocracy is an independent international news outlet based in the UK. Established in 2001, its reporting aims to inspire democratic debate and engagement across the world. Most of the funding OpenDemocracy receives comes from grants, with additional donations by individuals. The website lists the foundations and non-profit organisations that have donated money in the past, including the sums of the donations.

***Prospect* magazine:** www.prospectmagazine.co.uk/

Currently edited by the highly respected Alan Rusbridger, former *Guardian* editor, *Prospect* magazine is a monthly current affairs publication, based in the UK. Started in 1995 by a former *FT* journalist (among others), it is non-partisan and seeks to publish in-depth and far-sighted pieces. It is owned and supported by the Resolution Group as part of its not-for-profit, public interest activities. Notable articles include its work on the phone hacking scandal and the demise of the BBC.

The *Economist*: www.economist.com/

Founded in 1843 to further the cause of free trade, to my mind the *Economist* suffers from being wedded to the same outdated economic narrative as the *Financial Times*. It can, however, still produce worthwhile journalism. It has an international focus and covers a wide range of topics. Moreover, it prides itself on its editorial independence, and its constitution does not permit any individual or organisation to gain a majority shareholding.

ature *Private Eye*: www.private-eye.co.uk/

Private Eye's content is a unique blend of investigative journalism, satire and humour. Its signature approach uses wit and parody to expose hypocrisy and corruption in politics, business and media. It played a significant role in revealing the Post Office scandal. Beyond investigative pieces, the magazine features regular columns with a satirical and critical lens on current affairs. Unlike traditional newspapers, *Private Eye* is a cooperative owned by its journalists, ensuring editorial independence and a distinctive voice. Funding comes primarily from newsstand sales and subscriptions, allowing the magazine to operate free from advertising influences.

Fact-Checking and Whistle-Blowing

Full Fact: https://fullfact.org/

Full Fact checks claims made by UK politicians, journalists, public institutions and viral online content. It invites those responsible for errors to correct the record and reports on how they respond. You can, for example, search for your MP on Full Fact and very quickly get a sense of their integrity. The home page reads: '*Full Fact fights bad information. Bad information ruins lives. It promotes hate, damages people's health, and hurts democracy. You deserve better.*' It is somewhat dry and sticks to factual errors rather than covering the full taxonomy of deceit, but scores full marks for integrity, transparency, impartiality and rigour. It is a charity, funded transparently by a mix of businesses and charities, including Baillie Gifford, Google, Facebook, WhatsApp and the Joseph Rowntree Foundation.

DeSmog: www.desmog.com

As well as producing investigative journal articles on a variety of topics, from sustainability to policy, transport and social justice, DeSmog maintains six databases on Climate Disinformation, the 'Koch Network' of bogus think tanks funded by the billionaire Charles Koch, Air Pollution Lobbying (exposing businesses largely), Agribusiness, Industrial Aquaculture, and Advertising & Public Relations. There they have a record of individuals and organisations who have been guilty of activities such as climate disinformation. Usefully, clicking on one of these organisations or links gives a description and a list of other involved parties/people. Internationally focused, but with some UK-specific articles.

Global Witness: www.globalwitness.org/en/

Global Witness is a prominent international NGO known for its pursuit of environmental and human rights justice. Their work focuses on critical issues such as deforestation, the illegal wildlife trade, and the murder of land and environmental defenders. Global Witness aims to amplify the voices of those most affected by environmental destruction and human rights abuses, while advocating for systemic change and holding powerful actors accountable.

TheyWorkForYou: www.theyworkforyou.com/

Designed to empower UK citizens, TheyWorkForYou is a volunteer-run website launched in 2004. It functions as a tool for tracking the activities of Members of the UK Parliament, the Scottish Parliament, the Senedd and the Northern Ireland Assembly. It offers a user-friendly platform where you can easily search for your MP, track their voting record, follow their speeches and even read their expenses reports. This readily available information fosters transparency and enables citizens to hold their elected officials accountable. (I urge great caution in simplistically judging politicians by their voting track record because the reasons behind each vote can often be more subtle than is readily apparent.)

Tortoise Media Westminster Accounts page: www.tortoiseme dia.com/westminster-accounts-explore/

This gives you an instant picture of how any UK MP is externally funded. With easy-to-understand graphics, you can see how much income outside their salary an MP is getting and where from. This page is kept up to date in collaboration with Sky News. Tortoise Media do not reveal their funding partners on their website, but they do say this: 'Our partners, of course, know that we are a journalistic enterprise. Our independence is non-negotiable. If we ever have to choose between the partner and the story, we'll always choose the story.'

Led By Donkeys: https://twitter.com/bydonkeys

A political campaign group that began as an anti-Brexit initiative, highlighting the broken promises, lies and misinformation of pro-Brexit politicians by displaying their tweets and statements on billboards. The group later launched a crowdfunding campaign to legally acquire advertising space for their efforts. They continue to spotlight political scandals and hold individuals and the government accountable, frequently employing creative and satirical publicity stunts.

Political-Lies.com: https://political-lies.com

Formerly the boris-johnson-lies website, this is Peter Oborne's impressive and jaw-dropping catalogue of literally hundreds of lies and other uncorrected false and misleading statements by Boris Johnson and more recently also by other ministers and MPs. Not always up to date – but there has been a lot to keep up with. Careful and robust.

Bellingcat: www.bellingcat.com/

This Netherlands-based investigative journalism platform focuses on warzones, human rights abuses and crimes. Key cases include the Skripal and Navalny poisonings, wars in Ukraine and Syria, and the Malaysian Airlines Flight MH17. Funded mainly by donations from individuals and businesses.

Drilled: https://drilled.media/

A multi-media investigative journalism platform known for its critical examinations of the fossil fuel industry and its connections to governments and media outlets. It was set up by climate journalist Amy Westervelt in 2017. Drilled receives most of its funding through grants from environmental and social justice foundations, which are listed on its website, and through advertisements on its podcasts.

FactCheck: http://FactCheck.org

A US-based, non-partisan, non-profit 'consumer advocate' for voters that aims to reduce the level of deception and confusion in US politics. It monitors the factual accuracy of what is said by major US political players in the form of TV ads, debates, speeches, interviews and news releases. Funded by various foundations, by some businesses without editorial influence and by individuals, and never by unions, partisan organisations or advocacy groups. The funding is 100 per cent transparent.

Media and Journalism Research Center: https://journalismresearch.org/

The international Media and Journalism Research Center publishes articles and reports on media funding, influence and policy. The centre conducts research on various topics including media sustainability, state media and misinformation.

Media Reform Coalition: www.mediareform.org.uk/

The Media Reform Coalition is dedicated to advocating for a more democratic media system in the UK. Recognising the deficiencies of the current media landscape, they strive to address its shortcomings and promote the public interest. By challenging media corporations and fostering the development of an independent media ecosystem, they aim to enhance media plurality and accountability.

News Literacy Project: www.newslit.org

A non-profit with the stated aim of strengthening democracy including through helping the public to critically evaluate news. Website headlines: 'You have the power to stop misinformation. Use it.' Donors, however, include News Corp, NBC Universal News Group, the *New York Times*, Apple and Microsoft, some of which ring obvious alarm bells.

Politifact: www.politifact.com

Not-for-profit US national news organisation looking at specific statements made by politicians and rating them for accuracy. Has been accused of being more likely to call out Republican lies than Democrat lies, although the counter-argument is that there may be more Republican lies to call out. Funders include Facebook and Tiktok. Discloses all grants over $1,000.

Punditfact: www.politifact.com/punditfact/

An offshoot of Politifact, focused on checking the accuracy of claims by pundits, columnists, bloggers, political analysts, talk-show hosts and guests, and other members of the media.

Retraction Watch: https://retractionwatch.com/

A database and blog reporting on retractions of scientific papers, to make it harder for bogus research to get traction.

The Governance Project: www.ukgovernanceproject.co.uk/our-report/

Headed by Dominic Grieve KC, this has produced an interesting and detailed report on ethics, codes of conduct, parliamentary reform, the Civil Service, elections and democracy.

Transparency International: www.transparency.org/en

Dedicated to fighting corruption around the world, Transparency International (TI) does in-depth research on the causes and perceptions of corruption and has developed tools like the Corruption Perceptions Index (CPI) and the Global Corruption Barometer (GCB) to inform anti-corruption action and policies. Data from their research tools is freely available. Besides corruption at the national and multi-country levels, they also look at business corruption and campaign for more anti-corruption regulations globally. TI was founded in 1993 by former World Bank employees, led by Peter Eigen. Eigen states that his experience witnessing the impact of corruption during his work in East Africa inspired him to start the organisation. Most of TI's funding comes from government agencies and multi-lateral institutions. However, they also accept funding from corporate donors and have in the past accepted donations from companies themselves convicted of corruption offences (e.g., in 2014, a US $3 million contribution from Siemens, who in 2008 paid one of the largest corporate corruption fines in history). In January 2017, TI's Board of Directors disaccredited its US chapter, citing differences in philosophy. The US chapter faced accusations of having become a corporate front, heavily funded by multinational corporations such as Bechtel, Google, Pfizer and ExxonMobil, which raised concerns about its objectivity and alignment with TI's core mission.

Campaigning Organisations

There are many groups you can get involved with as a volunteer, and many people feel this is the most effective way of relieving climate anxiety – to take action. Here are just a few, focused on climate, democracy and truth, but I haven't even attempted to include the full breadth of fabulous environmental and social NGOs that you could get involved with.

Extinction Rebellion: https://rebellion.global/

A global movement for action on the climate emergency through non-violent civil disobedience. In 2019 their London protests were, to my mind, one of the most effective and perhaps the most inspiring interventions that the world has ever seen for the systemic changes we need. They are passionate and coherent in their calls for respect for people, the environment and the truth – values that I too have called for in this book.

Just Stop Oil: https://juststopoil.org/

Taking things a step further than Extinction Rebellion, Just Stop Oil garners more controversy, but also more press attention, which is part of its tactics – to fight back against the fossil fuel companies and campaign for no new oil and gas licences to be issued. Highly contentious as to whether their actions are useful or counterproductive. The people I know who have been involved are all supremely well intentioned. For me, JSO lacks the overall flavour of positivity and warmth that was so crucial to XR's best moments.

The Climate Majority Project: https://climatemajorityproject.com/

Founded by Rupert Read, this UK-focused movement is dedicated to supporting and uniting climate action groups nationwide. It aims to inspire a climate-concerned majority of people to mobilise and take action in their local communities and provides guidance on effective messaging, engagement and initiatives. Additionally, the project amplifies existing climate efforts while advocating for government policies to combat the climate emergency.

Climate Justice Alliance: https://climatejusticealliance.org/

CJA is an international coalition of grassroots movements dedicated to environmental and climate justice. Their mission is to build a regenerative economy by replacing current extractive systems, which are heavily based on fossil fuel extraction. Climate justice refers to addressing the disproportionate impact of climate change on marginalised communities and ensuring that the benefits of a low-carbon economy are equitably shared. CJA's campaigns advocate for a Just Transition, which emphasises local control of resources, sustainable practices and the empowerment of frontline communities most affected by environmental degradation.

MP Watch: www.mpwatch.org/

This UK-based, non-partisan, not-for-profit group seeks to improve and uphold standards in public life. It has a climate focus, and in the run-up to the 2024 election played an important role in raising awareness in their constituencies of the climate misdemeanours of some of the UK's worst politicians. You can get involved by supporting them with a donation or volunteering to be in one of their constituency groups.

Involve: www.involve.org.uk/

Involve is a non-profit dedicated to enhancing public participation in democracy. They provide resources and information to encourage citizen engagement, particularly in decisions related to achieving net zero and addressing the climate emergency.

Unlock Democracy: https://unlockdemocracy.org.uk/

Unlock Democracy is a campaign group advocating for UK electoral reform, believing that the current centralised political system impedes the country's ability to tackle long-term issues. Their 'Power to the People?' report outlines proposals to strengthen democracy and make voting in elections fairer and more accessible.

There are also many local climate/social action groups. Find out what is in your area and get involved.

Books

The challenge of picking the very best books out there is beyond daunting, but here is a list of some good ones that is neither definitive nor comprehensive.

Systemic Approaches
Scale, Geoffrey West (W&N, 2018)

Physicist Geoffrey West examines the fundamental principles that govern the growth and dynamics of living organisms, cities and companies. Through a detailed exploration of the mathematical patterns underlying these systems, West provides insights into the universal laws that shape their development and longevity. This work presents a rigorous analysis of the interconnectedness of biological and social structures, offering a comprehensive understanding of the complexities of growth and sustainability.

The Web of Meaning, Jeremy Lent (Profile, 2021)

This is a powerful blend of wisdom from every discipline you can think of, from the ancient to the modern, to arrive at a hopeful manifesto for the next evolution of human culture: an ecological and regenerative civilisation. This book takes some digesting, but it is worth it.

There Is No Planet B, Mike Berners-Lee (Cambridge University Press, 2021)

I wrote this book (updated in 2021) to provide a tour of the multidimensional Anthropocene challenge in one accessible read. In it, I look at the technical challenges and solutions, the social drivers and barriers, and the core values we need to adopt to both survive and thrive in the future. In some ways, it is the prequel to this book, with more emphasis on physical elements of the transition.

Truth and Politics

This chilling list of riveting and convincing accounts of the demise of truth and democracy and the inadequacy of decision-making processes in the UK also contains recipes for improvement.

The Rise of Political Lying, Peter Oborne (Simon & Schuster, 2005)

Oborne's first of two books cataloguing the rise of political dishonesty in the UK focuses largely on Blair, beginning with lies on his CV before being selected as a Labour parliamentary candidate, and on the early years of the Labour government. Disgusted though Oborne was at what he saw going on back then, it can be seen as just paving the way for the horrors to come over the next two decades.

The Assault on Truth, Peter Oborne (Simon & Schuster, 2021)

This short book, Oborne's sequel to *The Rise of Political Lying*, is chilling and meticulously evidenced. 'Britain and the West and everything we claim to stand for is under greater threat than at any time since the 1930s.' This is essential reading for anyone wanting to understand how bad things have become and how critical it is that we change course on political standards.

Bullingdon Club Britain: The Ransacking of a Nation, Sam Bright (Byline Books, 2023)

An account of how the very rich and privileged have gained corrosive power in the UK, from one of the co-founders of DeSmog.

Merchants of Doubt: How a Handful of Scientists Obscured the Truth on Issues from Tobacco Smoke to Global Warming, Naomi Oreskes and Erik M. Conway (Bloomsbury, 2012)

This carefully researched book unpicks how a small, loosely affiliated group of dishonest 'scientists' have cast doubt on established scientific consensus on tobacco smoke, acid rain, DDT, the hole in the ozone layer and, most devastatingly of all, global warming.

Lying, Sam Harris (Four Elephants Press, 2013)

A short gem of a book making the case that there is no such thing as a white lie, even to avoid embarrassment or hurt between friends and family, because the gain in trust and depth of relationship offered by uncompromising truth outweighs the short-term discomfort.

Post-Truth: How Bullshit Conquered the World, James Ball (Biteback Publishing, 2017)

Post-Truth dives into the rise of the post-truth era, where facts and evidence hold less weight than emotions and personal beliefs. Author James Ball, an award-winning journalist, argues that this phenomenon isn't just about fake news and social media. It's about a much larger shift in our political, media and online landscapes that has led to a devaluation of truth.

Post Truth: The New War on Truth and How to Fight Back, Matthew d'Ancona (Ebury Press, 2017)

In *Post Truth*, Matthew d'Ancona argues that prioritising feelings over facts and misinformation spreading through social media have eroded trust in facts. He proposes solutions like quality journalism to combat this 'war on truth'.

Post-Truth, Lee McIntyre (MIT Press, 2018)

Yes, a third book with this title. Lee McIntyre explores the rise of the post-truth era, where ideology trumps objective facts. The book explores how social media, cognitive biases and a decline in trust in traditional media contribute to this phenomenon.

Code of Conduct: Why We Need to Fix Parliament – and How to Do It, Chris Bryant MP (Bloomsbury, 2023)

MP Chris Bryant delves into the erosion of standards in British politics, exposing conflicts of interest and misconduct, and examining their impact on public trust, while proposing reforms to restore integrity and accountability in Parliament.

Why We Get the Wrong Politicians, Isabel Hardman (Atlantic Books, 2018)

Investigative journalist Isabel Hardman explores why voters feel stuck with politicians they distrust, and why the system makes it so hard for the best people for the job to find their way into political life and survive once they are there. Hardman looks at both flaws in the system and potential paths to a more representative democracy.

Downward Spiral, John Bowers KC (Manchester University Press, 2024)

John Bowers KC charts the collapsing public standards in the run-up to the Johnson era, through Johnson's time in office and beyond.

Overruled, Sam Fowles (Simon & Schuster, 2023)

A look at the UK's crumbling democracy through the lens of eight legal cases that Fowles was involved in, including the proroguing of Parliament.

Politics on the Edge, Rory Stewart (Jonathan Cape, 2023)

Rory Stewart's account of his 10 years as a Conservative MP will leave you screaming for change: in standards on public life, in the mechanics of the political system, in the processes by which political decisions are made in the UK, and for a change in the people running the country from most of those he worked with.

But What Can I Do? Alastair Campbell (Penguin, 2024)

Alastair Campbell's book seeks to empower readers disillusioned with politics by analysing the factors contributing to its current disarray, such as governmental dishonesty and the erosion of democratic values. Drawing on Campbell's extensive experience, it provides practical tips on campaigning, supplying readers with the tools and motivation to effect positive change and become active political participants.

Economics, Inequality and Society

Limits to Growth, Donella Meadows, Dennis Meadows, William Behrens and Jorgen Randers (Universe Books, 1972)

A seminal book which sounded the alarm on the madness of the constant growth model over 50 years ago. Along with *Silent Spring*, these two books alone demonstrate that the need for humans to change our relationship with the environment and rethink our economics, in response to our arrival in the Anthropocene, has been clear for many decades. When the evidence has been so clear for so long, the subsequent question has become 'Why haven't we heard their warnings?'

Doughnut Economics, Kate Raworth (Chelsea Green, 2017)

Perhaps the clearest argument yet for an Anthropocene-fit economic system. This book provides a workable framework for businesses and governments to move beyond outdated growth- and GDP-focused models and into a better way of thinking for a thriving future.

The Spirit Level, Richard G. Wilkinson and Kate Pickett (Allen Lane, 2009)

Drawing on extensive research, and packed with lots of graphs and data, Wilkinson and Pickett's book argues that more equal societies experience lower rates of violence, crime, mental illness and addiction, and an overall higher quality of life. They explore the reasons behind this link and propose policies that can promote a fairer and healthier society for all. Although it's quite an old book now, and inequality has worsened since it was written, the principles remain.

Poverty, by America, Matthew Desmond (Allen Lane, 2023)

Following on from his excellent, Pulitzer Prize-winning book *Evicted* (2017), sociologist Matthew Desmond analyses the deep inequalities in American society (and globally). He points to the fact that one-third of the world's wealth is sitting in offshore tax havens, while the poor struggle to feed themselves and afford basic utilities.

The Price of Inequality, Joseph Stiglitz (W. W. Norton & Company, 2012)

Nobel laureate economist Joseph Stiglitz exposes the stark reality of wealth disparity in the US and its detrimental effects on the entire economy. He delves into the root causes of inequality, including globalisation, deregulation and tax policies that favour the wealthy. Stiglitz demonstrates how these widen the gap between the rich and the poor, hindering economic growth, social mobility and, ultimately, the wellbeing of a nation.

Why We Can't Afford the Rich, Andrew Sayer (Policy Press, 2015)

Social scientist Andrew Sayer challenges the myth that the wealthy are essential wealth creators. He exposes the mechanisms that allow the top 1 per cent to accumulate vast fortunes at the expense of the majority.

Prosperity Without Growth, Tim Jackson (Routledge, 2017)

Tim Jackson challenges the conventional notion that endless economic growth is both desirable and feasible. Drawing on economic theory, environmental science and social philosophy, Jackson presents a compelling case for redefining prosperity beyond GDP growth. He argues for a sustainable economic model that prioritises wellbeing, social equity and environmental stewardship.

Humankind: A Hopeful History, Rutger Bregman (Bloomsbury, 2021)

If you need some uplifting reading to counter the neoliberal narrative that humans are innately greedy, this is the book for you. Rutger Bregman makes a compelling argument that we are not all in it for ourselves; in fact, we are a collaborative species, and there is plenty of evidence that we can pull together and get through the challenges ahead.

Citizens, Jon Alexander, Ariane Conrad and Brian Eno (Simon & Schuster, 2022)

Drawing on diverse perspectives and real-world examples, this book inspires readers to become active participants in building a more just, equitable and democratic world.

The Lorax, Dr Seuss (Random House, 1971)

If only we'd all grown up with this children's book and absorbed its key messages. Brilliantly sets out the case for respecting planetary boundaries with fabulous illustrations and poetic wisdom. Perhaps it should be on the national curriculum for all small children?

Environment, Ecology, Climate and Food

I know I have missed out so many great books here. Note that I have not included any of the climate techno-optimistic books that, as I have articulated earlier, are a form of climate denial and can so easily do more harm than good.

Silent Spring, Rachel Carson (Houghton Mifflin, 1962)

Rachel Carson's groundbreaking work, *Silent Spring*, exposed the devastating effects of pesticides on the environment. By blending scientific research with vivid storytelling, Carson's book is often credited with igniting the modern environmental justice movement, raising awareness about the interconnectedness of all living things and the urgent need for sustainable practices. The book remains a powerful call to action for protecting nature and all species living on Earth.

The Climate Book, Greta Thunberg (with many contributors) (Penguin Random House, 2022)

Over 100 esteemed authors (plus me) contribute a chapter each to cover the climate emergency from multiple angles. It is as close as there is to an encyclopaedia on the topic.

The New Climate War, Michael Mann (Scribe UK, 2021)

American climatologist and geophysicist Michael E. Mann exposes how fossil fuel companies have waged a decades-long campaign to deflect blame and responsibility for the climate crisis and delay action on addressing it. He outlines a plan for how we can take back control from the polluters and address the climate emergency.

Five Times Faster – Rethinking the Science, Economics and Diplomacy of Climate Change, Simon Sharpe (Cambridge University Press, 2023)

Sharpe sets out how we should rethink our strategies in the fields of science, diplomacy and economics to make the rapid changes we need to tackle climate breakdown.

How Bad Are Bananas? The Carbon Footprint of Everything, Mike Berners-Lee (Profile Books, 2020)

I can't resist mentioning my own book. I wrote the first edition in 2010 and gave it a complete overhaul in 2020, to update the numbers and include new items, but most importantly to better reflect the unfolding emergency and my latest thoughts on how we should approach it. It aims to be both light and deadly serious, accessible and realistic. Items that look trivial are usually included because they nod to bigger issues – but occasionally are just there for fun. Carbon numbers can be essential for perspective, but often their most important role is as a gateway into bigger discussions.

Power Up, Yasmin Ali (Hodder Press, 2024)

This is an excellent insider's tour of existing global energy infrastructure, showing both the challenges and opportunities of scaling-up various renewable technologies. It allows the reader to put things into perspective and see what's really going on behind the scenes, plus brush up their knowledge of the history of different technologies (e.g. first commercial wind turbine).

FOOD TRANSITION

Ravenous, Henry Dimbleby (Profile Books, 2024)

Dimbleby's *Ravenous* is based on the excellent *National Food Strategy* report that he led, which the UK government at the time commissioned but then largely ignored. It exposes the flaws in our modern food system, which often seems to prioritise profit over health and sustainability. This broken system contributes to obesity, environmental degradation, and a growing disconnect from where our food comes from. Dimbleby goes on to offer a roadmap to a healthier future, urging the reader to reassess their relationship with food and advocating for reforming the food system in a way that benefits individuals and the planet alike. If only the UK government had adopted every one of his recommendations.

Regenesis, George Monbiot (Penguin, 2023)

Probably the best articulation of the case for a fully vegan food system. An enlightening read even if, in my view, it goes one step further than the evidence.

Feeding Britain, Tim Lang (Pelican Books, 2021)

Drawing on his expertise in food policy, Lang examines the challenges of ensuring food security for all citizens while navigating the impacts of the climate emergency and socio-economic disparities. He makes the case that the UK's food supply is less secure than many assume, and he addresses issues of accessibility, sustainability and health.

Wild Fell, Lee Schofield (Penguin, 2023)

Lee Schofield, former site manager at RSPB Haweswater in the Lake District, offers a compelling narrative of his decade-long journey rewilding two hill farms and their expansive upland habitat. This passionate account details pioneering conservation efforts amidst a diverse landscape of woodlands, bogs, mountains and meadows, while navigating the delicate balance with local farming traditions.

Online Blogs/Channels/Podcasts
The Great Simplification: www.thegreatsimplification.com/episodes

Nate Hagens interviews leading scientists on a wide range of subjects. The high-quality discourse makes this essential listening.

The Crucial Years: https://billmckibben.substack.com/

With decades of experience as a climate activist, author and 350.org founder, Bill McKibben uses his blog to provide insightful commentary on the urgency of addressing the climate emergency, highlighting key issues and showcasing grassroots efforts.

The News Agents: https://tinyurl.com/news-agents

Emily Maitlis, Jon Sopel and Lewis Goodall. Three top former BBC journalists make a great and accessible contribution to political analysis. I wish they were more careful in their choice of adverts, which often push climate-crazy companies.

The Rest Is Politics: www.youtube.com/@restispolitics

Alastair Campbell and Rory Stewart's discussions on everything under the sun, in which they 'agree to disagree agreeably', make a good contribution to thoughtful analysis of global and UK politics. That doesn't mean I always agree with them, and sometimes, to my mind, they are woefully off the ball when it comes to the Polycrisis. Their interview on *The Rest Is Politics: Leading* with Bill Gates had me wanting to tear my hair out. Nevertheless, I often value their insights. There is now an offshoot, *The Rest Is Politics: US*, which is also good.

Gary's Economics, YouTube channel: www.youtube.com/@garyseconomics

Author of *The Trading Game* (2024), former trader Gary Stevenson made millions of pounds betting that inequality would destroy the economy. He left trading to campaign against inequality and to raise awareness for everyday people about what is happening economically in the UK and beyond.

Outrage and Optimism: www.outrageandoptimism.org/

Discusses the climate emergency with a mix of urgency, passion and hope. Hosted by Christiana Figueres, Tom Rivett-Carnac and Paul Dickinson, this podcast features engaging discussions, interviews and insights from thought leaders and activists around the world. The hosts also produced a special series on the future of food, sponsored by the Ikea Foundation.

Films/Documentaries

Don't Look Up – 2021 star-studded and chillingly close-to-the-bone allegory of the human response to climate change. Written, co-produced and directed by Adam McKay.

The Century of the Self – 2002 four-part documentary by Adam Curtis which gives a thought-provoking insight into consumerism, advertising and subtle social control. It explains a lot about the world we live in.

Plastic Fantastic – 2023 documentary about the impact of plastic on both planetary and human health.

Once You Know – 2020 documentary by Emmanuel Cappellin examining how we can and ought to cope with the changes that will be brought about by climate breakdown.

Before the Flood – 2016 documentary in which Leonardo DiCaprio meets with scientists, activists and politicians to examine the dangers of climate change and possible solutions.

An Inconvenient Truth – 2006 documentary featuring Al Gore's relentless efforts to try to get the world to wake up to the seriousness of our changing climate.

An Inconvenient Sequel: Truth to Power – 2017 follow-up to his earlier film in which Al Gore fearlessly demands action to ensure a survivable future for the next generation.

Climate Change: The Facts – 2019 documentary by the BBC finally saying it as it is on climate change, or as it was seen to be back in those days. Sadly, things have got worse, and it is now out of date.

Financial Information

Banking

Put your money with a bank that cares about your future. You can find advice and information at:

Bank.Green: https://bank.green/

Bank.Green offers a tool designed to empower individuals to make informed decisions about their banking choices. With the ability to check whether your bank is investing in fossil fuels across over 60 countries, this platform enables users to align their finances with their values.

Banking on Climate Chaos: www.bankingonclimatechaos.org/

Banking on Climate Chaos publishes an annual report examining the intricate ties between the finance sector and the fossil fuel industry. Delving into the financial commitments of the world's largest 60 banks to fossil fuel expansion, the report sheds light on the detrimental impact of such financing on climate justice.

MotherTree: www.mymothertree.com

Helps businesses to find ethical banks to suit their needs. We used them in my business to help us with a long-overdue banking switch.

Pensions

Research ethical pension providers here:

Ethical Consumer: www.ethicalconsumer.org/money-finance/shopping-guide/ethical-pensions

This article is a guide to ethical pensions. It explains what makes a pension ethical and how to find one. Some ethical pensions avoid investing in certain sectors, such as fossil fuels or tobacco, while others focus on companies that are environmentally friendly or socially responsible.

Good With Money: https://good-with-money.com/2024/02/26/top-9-ethical-pension-funds/

This article is about ethical pension funds. It explains what they are and why you might choose one. The article also details several different ethical pension funds offered by various providers. Some of the important points to consider when choosing an ethical pension fund are the fees, the investment strategy and what is excluded.

Make My Money Matter: https://makemymoneymatter.co.uk/

Make My Money Matter is a UK campaign platform founded by filmmaker Richard Curtis and former Number 10 special adviser Jo Corlett, advocating for aligning personal finance and pensions with the climate emergency and social justice goals. They promote divestment from fossil fuels and investment in sustainable alternatives to drive positive change towards a low-carbon economy. Through educational resources, advocacy efforts and partnerships with organisations and influencers, Make My Money Matter seeks to empower individuals to make informed decisions about their finances.

ACKNOWLEDGEMENTS

So many people helped me to write this book.

Everyone at Small World Consulting has helped by chipping in their thoughts on content as well as endless cover designs and ideas for the title, not to mention politely tolerating my distraction from the day job. Mairéad Brown's help was completely invaluable: reading drafts, researching, finding references, suggesting phrases, encouraging and nudging me along. Jenny Lyon did wonderful work straightening out references. Mira Kracke created and tidied up images.

Tom Mayo did a skilful first edit on the earlier sections. Anna Gunstone helped with developing criteria for assessing people and media and later picked through the whole book. Thanks to all those who read and commented on sections, including John Bowers, Estelle Dehon, Pooran Desai, Judi Marshall, Bill McGuire, Kate Rawles, Kate Raworth, Jessica Townsend and Rebecca Willis. I was also grateful for discussions with many people but especially Peter Oborne, Caroline Lucas, Geoffrey West, Tim Farron, Jonathan Rowson, Rory Stewart, Alastair Campbell, Rupert Read and Amber Rudd.

Thanks to all those who have come along to my talks over the past five years, and in doing so helped me to work up and test out the ideas in this book.

Thanks to Matt Lloyd, my patient and helpful editor at Cambridge University Press, and also to Ruth Boyes, Susie Francis and Phil Meyler. Lindsay Nightingale has been an amazing copyeditor.

Most of all, thanks to wonderful Liz for continuous love and support through a difficult writing process.

NOTES

Introduction

1. *How Bad Are Bananas? The Carbon Footprint of Everything* (Profile Books, 2020). Published in 2010, but then when people were still buying it 10 years later, I completely revised it (with help from some colleagues), which was harder work than writing the original, because the pressure to be robust was greater than ever.
2. For example, the *Guardian* ran a major 'Keep it in the Ground' campaign over many months, the *Economist* ran a front cover headline 'The burning question', with a graphic almost identical to that of our book cover (it felt cheeky but we welcomed the message): https://tinyurl.com/econ-burn-q, and Bill McKibben was campaigning to great effect.
3. COP stands for 'Conference of the Parties', which since 1995 has served as the annual formal meeting of the United Nations Framework Convention on Climate Change to assess global progress on tackling climate issues. After nearly three decades of COPs, emissions are still rising.
4. There has always been a bit of noise on the curve – ups and downs between individual years. Since 2020 the trend has looked a little better, but there is no statistical significance in this. Whether it is the temporary effect of the pandemic, or war, or the very first glimmers of a deliberate response to the climate emergency bearing fruit, remains too early to say.

1 The Challenge Ahead

1. Donella Meadows, Dennis Meadows, William Behrens and Jorgen Randers, *Limits to Growth* (Universe Books, 1972); and

Rachel Carson, *Silent Spring* (Houghton Mifflin, 1962). See also Appendix 4, Books section.

2 Standing Further Back

1. I and others have written about this at length elsewhere (see *The Burning Question*, *How Bad Are Bananas?* and *There Is No Planet B* (all referenced in Appendix 4).
2. For example, on 28 February 2024, the UK Climate Change Committee's letter of advice to government on the Third Carbon Budget carry-over contained the following note: '**Domestic transport**: The projected fall in annual emissions of approximately 33 MtCO$_2$e was driven by assumed improvements to conventional vehicle efficiencies. Actual emissions fell by less than half this amount with the underachievement largely due to a shift to larger cars. In addition, the distance driven by cars was slightly higher (by 5%), and new-car CO$_2$ emissions were higher (by 19%).' https://tinyurl.com/CCCletter3CB-carryover
3. See 'There's no need to lose our minds over the Jevons paradox', *Financial Times*, Tim Harford, 17 May 2024 (https://tinyurl.com/FT-JevonsParadox) and my comment on this article (sorry about the paywall). Harford cites Hannah Ritchie's book *Not the End of the World* (Vintage, 2024) in which she makes the case that reduction in consumption-based emissions in some countries demonstrates absolute decoupling of emissions from energy use and GDP (in other words, saying energy use can grow while emissions fall). The problem is that the interaction with the rest of the global economy is not fully accounted for, even by the use of consumption-based emissions metrics (i.e. metrics that include net embodied emissions imported in goods and services). For example, consumption-based metrics don't account for the tendency of a coal exporter to switch its market, nor for GDP growth to shift into other countries.

4. United Nations Climate Change '2023 NDC Synthesis Report': https://tinyurl.com/UNFCCC-Proj-GHG-Emissn-levels
5. In 2013 in *The Burning Question* (Profile Books), Duncan Clark and I described this as the 'balloon squeezing effect'. See Part 2, 'Squeezing the balloon'.
6. Geoffrey West's book *Scale* (Weidenfeld and Nicholson, 2018) is a fascinating and recommended read. See also his TED talk: https://tinyurl.com/gWest-surprising-math-cities
7. West, B.J. (2020). 'Sir Isaac Newton stranger in a strange land.' *Entropy*, 22, 1204.
8. Here's a very simple model using a bit of calculus, illustrating for maths nerds why the ¾ ratio between mass and metabolic rate that West found in his data might not be totally surprising. Let's approximate that when a tree or an animal grows, it stays the same shape, and also assume that the liquid and nutrition transport requirement is indicative of the metabolic rate required to live. Then the mass goes up with the cube of the height – the third power. But the transport distances also go up in proportion to the height, so if the metabolic rate stayed the same per kilogram, the total transport effort would have to go up with the fourth power of the height. To get the proportionate change in mass and transport requirements per unit increase in height, divide the differential of the mass by the mass and the differential of the transport requirement by the transport requirement. The fact that one goes with the third power of the height and the other with the fourth power of the height leads to a ¾ ratio between the two. So, when mass goes up by 100%, the other only goes up by 75%: $(dm/dh)/m = ¾ \times (dt/dh)/t$ where h is the height, m is the mass and t is the transport requirement. The difference between the 100% and 75% growth is the efficiency improvement.
9. Lupi, T.M., Nogales, S., León, J.M., Barba, C. and Delgado, J.V. (2015). 'Characterization of commercial and biological growth curves in the Segureña sheep breed.' *Animal*, 9, 1341–1348.

10. It came to light in the UK's review into the government's response to the COVID-19 pandemic that some senior ministers struggled with the concept of an exponential curve. It is an understatement to say that this is important for those in charge of containing a pandemic. Patrick Vallance, the UK's Chief Scientific Adviser in the pandemic, recalled being on a group call with scientific advisers from various countries, when one said their leader could not understand exponential curves, 'and the entire phone call burst into laughter because it was true in every country'). His diaries also recount that Boris Johnson found it a 'real struggle' to understand some graphs. The *Guardian*, Peter Walker, 20 November 2023, 'What we learned from Patrick Vallance at the Covid inquiry'. https://tinyurl.com/Guardian-PVallance-CovidEnq
11. Steffen, W., Broadgate, W., Deutsch, L., Gaffney, O. and Ludwig, C. (2015). 'The trajectory of the Anthropocene: The Great Acceleration.' *The Anthropocene Review*, 2, 81–98. https://tinyurl.com/trajectory-anthrop
12. Graph use permission from Geoffrey West. There is also an interesting *FT* article discussing it further: *Financial Times*, Izabella Kaminska, 9 June 2017, 'Why things stop growing and why other things don't'. https://tinyurl.com/West-podcast-thgsstopgrowing

3 The Outer Layer of the Polycrisis

1. Supran, G. et al. (2023). 'Assessing ExxonMobil's global warming projections.' *Science*, 379, eabk0063, and also written up accessibly by Alice McCarthy in the *Harvard Gazette*, 12 January 2023, 'Exxon disputed climate findings for years. Its scientists knew better'. https://tinyurl.com/Harvd-Exxon-scientists-knew

    Also, the BBC produced a compelling documentary interviewing Exxon's own scientists: 'Big Oil v the World', 2022. https://tinyurl.com/bbc-big-oil-v-world

2. Duncan Clark and I wrote about this in detail in chapter 1 of our 2013 book *The Burning Question*. This in turn drew upon Andrew Jarvis et al. (2012), 'Climate–society feedbacks and the avoidance of dangerous climate change.' *Nature Climate Change*, 2, 668–671, which found that $U = e^{a(t - t1)}$ where U is anthropogenic CO_2 emissions, t is the year, $t1$ is the year 1883, and 'a' turns out to be 0.0179 +/– 0.0008. In other words, emissions have been rising at 1.8 per cent per year, with very little deviation from the exponential curve (www.bitly.com/carbon-curve). In the years since 2012, the rise in emissions has looked more linear – which you can call good news if your glasses are sufficiently rose-tinted.
3. Global Carbon Project, 2024.
4. Data from Wikipedia, https://en.wikipedia.org/wiki/Methane_emissions
5. Alongside GWP100 and GWP20, GWP* is another metric, created by Myles Allen and others at Oxford University in an attempt to bring methane and CO_2 into one metric while taking account of their different characteristics. It is useful if properly used, but if misinterpreted can be misused to create an impression that major producers of unnecessary amounts of methane are environmentally wonderful if they make even small reductions. We wrote a briefing paper on this, and in turn it draws upon the key academic papers underpinning GWP*, as well as giving more background to GWP100 and GWP20. Dmitry Yumashev, *GWP*: Applications and Misapplications* (Small World Consulting, 2024). www.sw-consulting.co.uk/gwpstar
6. CarbonBrief, Ayesha Tandon, 20 March 2023, 'Methane emissions from wetlands have risen faster this century than in even the most pessimistic climate scenarios, new research finds'. https://tinyurl.com/surge-methane-wetlands

   See also Zhang, Z. et al. (2023). 'Recent intensification of wetland methane feedback.' *Nature Climate Change*, 13, 430–433. https://tinyurl.com/wetland-methane-fdbk

7. National Oceanic and Atmospheric Administration (NOAA). 'Increase in atmospheric methane emissions set another record during 2021', April 2022. https://tinyurl.com/NOAA-recordmethane2021
8. 'The 2023 Annual Climate Summary', Copernicus Climate Change Service. https://climate.copernicus.eu/global-climate-highlights-2023
9. See the reference in endnote 8 above, section 3: 'Was the unusual warmth of 2023 expected?' 'The change in annual temperature from 2022 to 2023 was larger than any change from one year to the next in the ERA5 data record. 2023 is also unusual in that its record temperature is for a year in which the El Niño was building up rather than declining.'
10. Average temperature change across all the Earth's land and sea area is the best measure we have of global heating. Land tends to warm much more than sea, so, on average, humans experience much more than stated global temperature change. There are also regional variations, with some places getting colder and others much hotter.
11. Copernicus, 4 March 2024, 'February 2024 was globally the warmest on record – global sea surface temperatures at record high'. https://tinyurl.com/Feb24-sea-surface-temp
12. International Energy Agency, March 2024, 'CO_2 emissions in 2023'. https://tinyurl.com/IEA-CO2-2023
13. Prevention Web, 16 January 2024, 'Ten extreme climate events of 2023'. www.preventionweb.net/news/ten-extreme-climate-events-2023
14. CNN, January 2024, 'Extreme cold kills more than 150 people in Afghanistan, Taliban says'. https://tinyurl.com/afghan-freezing-deaths
15. World Health Organization, 21 December 2023, 'Dengue – global situation'. https://tinyurl.com/WHO-dengue-global
16. Chartered Institute for Logistics and Transport, Fred McCague, 2023, 'Record year in the Northwest Passage as part of busy Arctic season'. https://tinyurl.com/busy-NW-passage-Arctic

17. Forbes, Daphne Ewing-Chow, December 2023, 'Here are the foods hit hardest by climate change in 2023'. https://tinyurl.com/foods-hit-climate-change
18. 'The potato industry in 2023 has been significantly affected by climate change, leading to unpredictable weather patterns and increasingly hostile growing conditions. These changes have posed challenges for potato growers worldwide, impacting yield and quality. Farmers contended with droughts, heatwaves, flooding in some regions, and unseasonal rains in others.' 'The evolving landscape of the global potato industry: trends, challenges and innovations in 2023', *Potato News Today*, January 2024. https://tinyurl.com/potato-trends-challges
19. Potsdam Institute for Climate Impact Research, 2024, 'Tipping elements – big risks in the Earth System'. Also Armstrong McKay, D., Abrams, J., Winkelmann, R. et al. (2022). 'Exceeding 1.5°C global warming could trigger multiple climate tipping points.' *Science*, 377, eabn7950. https://tinyurl.com/1pt5-multiple-tipping
20. Van Westen et al. (2024). 'Physics-based early warning signal shows that AMOC is on tipping course.' *Science Advances*, 10. https://tinyurl.com/early-warning-AMOC-tipping
21. Same reference as endnote 20, above: https://tinyurl.com/early-warning-AMOC-tipping
22. See article by Ritchie, P.D.L., Smith, G.S., Davis, K.J. et al. (2020). 'Shifts in national land use and food production in Great Britain after a climate tipping point.' *Nature Food*, 1, 76–83. https://tinyurl.com/climateshifts-UKfood-land-use. Stefan Rahmstorf, Professor of Physics of the Oceans and head of Earth System Analysis at the Potsdam Institute, describes the tipping point risk in a recommended YouTube video. He puts the rate of fall in Europe at 4 °C per decade, not 3 °C. https://tinyurl.com/tip-point-AMOC
23. Same reference as endnote 20, above: https://tinyurl.com/early-warning-AMOC-tipping

24. António Guterres, UN Secretary-General. Cited in the *Guardian*, Ajit Niranjan, 27 July 2023, '"Era of global boiling has arrived", says UN chief as July set to be hottest month on record'. https://tinyurl.com/era-global-boiling
25. 'We asked 380 climate scientists what they felt about the future'. Special report by Damian Carrington in *Guardian Weekly*, 17 May 2024. https://tinyurl.com/380scientists-despair. The *Guardian* has sent a survey to 843 IPCC lead authors since 2018 and received 380 responses. The mean temperature rise they predicted was 2.7 degrees.
26. IEA, 1 March 2024, 'CO_2 emissions in 2023: A new record high, but is there light at the end of the tunnel?' https://tinyurl.com/IEA-CO2-2023
27. Biography of British economist and demographer Thomas Malthus: www.britannica.com/money/Thomas-Malthus
28. Our World in Data graph showing global population growth since 1850 and forward to 2100: https://tinyurl.com/world-popul-byregion-1800-2100
29. According to Amy J. Lloyd, 'Education, literacy and the reading public', British Library Newspapers (Gale, 2007). https://tinyurl.com/educ-literacy-reading-public
30. European Commission (EU Science Hub), 2019, 'Educating all girls is key for global population size – EU Demographic Scenarios'. https://tinyurl.com/educate-girls-population
31. As demonstrated rather neatly using boxes, in this 2010 video of a TED talk by Hans Rosling: https://tinyurl.com/hRosling-pop-growth-boxes
32. This article by Ashley Ahn (NPR, 19 March 2023) discusses the very low fertility rate in South Korea: https://tinyurl.com/SKorea-lowest-fertil-rate
33. For example, Hannah Ritchie and Max Roser in their 2023 article, 'How has world population growth changed over time?' https://ourworldindata.org/population-growth-over-time

34. Three population studies in particular have attempted to include factors other than the UN's simple demographic projections (as reproduced, for example, by Our World in Data; see endnote 27 in this chapter). The *Lancet* factors in changes in healthcare and mortality rates, while Wittgenstein factors in education. These two studies each capture different aspects of the key influences on population growth, but they come up with similarly lower estimates than the UN. Earth For All attempted to capture all of these influences through the single link to GDP and, hey presto, in doing so its estimate is about twice as optimistic – compared with the UN's – as either Wittgenstein or the *Lancet*. None of these modelling approaches are at all perfect. After the pessimism of the fossil fuel emissions data, it is refreshing to see that sometimes the modelling can give us good news as well as bad. See the *Guardian*, Jonathan Watts, 27 March 2023, 'World "population bomb" may never go off as feared, finds study'. https://tinyurl.com/guardian-no-world-pop-bomb

35. United Nations World Population Prospects 2022. https://tinyurl.com/UN-world-pop-prospects-summary

    The *Lancet* and Institute for Health Metrics and Evaluation, 20 March 2024. https://tinyurl.com/Lancet-Fertility-Pop-Patterns

    Wittgenstein Centre: https://tinyurl.com/wittgenstein-data-explorer

    Earth For All, 27 March 2023. https://tinyurl.com/earth4all-pop-peak-below-9bn

36. *Birthgap*: an alarmist documentary that fails to consider the counter-arguments I have made. https://tinyurl.com/birthgap-vid

37. An example is this 37-minute episode of a podcast by 'The News Agents', entitled 'Is the world running out of babies?' (Global, 27 March 2024): https://tinyurl.com/running-out-of-babies

38. Same reference as endnote 37 above: https://tinyurl.com/running-out-of-babies

39. For more on this, try David Graeber's book *Bullshit Jobs* (Simon & Schuster, 2018). I also wrote a short piece about jobs in

There Is No Planet B, laying out three criteria for a job to be a good thing: (1) it contributes in some way to the net wellbeing of people and/or planet (directly or indirectly), (2) it is fulfilling for the person who does it, and (3) the associated remuneration helps to distribute wealth in a way that is consistent with everyone in the world having enough wealth for a high quality of life.

40. 'Even before the cost-of-living crisis, male farm workers are three times more likely to take their own lives than the male national average, and every week three people in the UK farming and agricultural industry die by suicide. Worryingly, we expect these figures to rise as the cost-of-living crisis exacerbates the longstanding challenges farmers already face – including higher living costs, known as the rural premium, increased rates of loneliness, and isolation.' *Agriland*, Eva Osborne-Sherlock, 11 September 2023, '3 people in agriculture die every week – BACP'. https://tinyurl.com/3-farm-suicides-a-week

    The US suicide rate among farmers is three and a half times the national average and grew by 48 per cent between 2000 and 2018. *New York Times*, 24 April 2023, 'A death in dairyland spurs a fight against a silent killer'. https://tinyurl.com/USagri-depression-suicide

41. 'Eating Better', February 2020. https://tinyurl.com/eat-better-chicken

42. The Wye & Usk Foundation, 8 June 2020. 'Nation's "favourite" river facing ecological disaster'. https://tinyurl.com/Wye-facing-ecological-disaster

43. Figure adapted from Our World in Data with long historical data from Barnosky (2008), 1900 data from Smil (2011), and 2015 data from Bar-On et al. (2018). https://ourworldindata.org/wild-mammals-birds-biomass

44. The original paper is: Berners-Lee, M., Kennelly, C., Watson, R. and Hewitt, C.N. (2018). 'Current global food production is sufficient to meet human nutritional needs in 2050 provided

there is radical societal adaptation.' *Elementa Science of the Anthropocene*, 6, 52. Since then, we've updated the numbers and responded to a change in FAO data that now distinguishes better between human-grade food and animal feed, some of which is still digestible by humans. The paper includes details on other nutrients, regional analysis and future scenarios. Some of the detail from the original paper is also in my book, *There Is No Planet B*.

45. World Health Organization and Food and Agriculture Organization of the United Nations, 2001, for the global average calorific intake required for healthy life (2,358 kcal per person per day): 'Human energy requirements', Report of a Joint FAO/WHO/UNU Expert Consultation. https://tinyurl.com/FAO-human-energy-reqts. In round numbers, the NHS says 2,500 kcal/p/d for men and 2,000 for women, on average. https://tinyurl.com/NHS-understanding-calories

46. *Le Monde*, Stéphane Mandard, 18 May 2022, 'Pollution is responsible for 9 million deaths each year worldwide'. https://tinyurl.com/pollution9m-global-deaths-yr

47. *Global Plastics Outlook: Economic Drivers, Environmental Impacts and Policy Options* (OECD, 2019). https://tinyurl.com/OECD-global-plastics-outlook

    Further plastics data from https://stats.oecd.org. I used OECD predictions to extrapolate from the baseline year: 460 million tonnes in 2019 to 500 million tonnes in 2024.

48. United Nations online exhibit 'Plastic is Forever', launched in 2021, claims that half of the 8.3 billion tonnes was created in the past 13 years. I'm assuming they have used data up to 2020 and have extrapolated from there using OECD production projections from 2019. www.un.org/en/exhibits/exhibit/in-images-plastic-forever

49. 3,150 million tonnes in 2019 and now probably closer to 3,500 million tonnes as I write this in 2024, for a global population of just over 8 billion. So more accurately 430 kg per person.

50. All plastic flow statistics are from *Global Plastics Outlook: Economic Drivers, Environmental Impacts and Policy Options*. https://tinyurl.com/OECD-plastics-keyfindings
51. *Politico*, 27 April 2024, 'Bird flu in US cows: should Europe be worried?' www.politico.eu/article/bird-flu-us-cows-europe-worry-who/
52. World Health Organization, November 2023, 'Antimicrobial resistance fact sheet'. https://tinyurl.com/WHO-antimicrob-resistance
53. The *Lancet*, 29 February 2024, 'Worldwide trends in underweight and obesity from 1990 to 2022: A pooled analysis of 3663 population-representative studies with 222 million children, adolescents, and adults'. https://tinyurl.com/Lancet-world-underwt-obesity
54. Gearhardt, A.N. et al. (2023). 'Social, clinical, and policy implications of ultra-processed food addiction.' *BMJ*, *383*, e075354. https://tinyurl.com/BMJ-implic-ultraproc-food. Also Tony Hicks's article in *Medical News Today*, 18 October 2023, 'Ultra-processed foods may be as addictive as smoking'. https://tinyurl.com/ultraproc-addictive-smoking
55. These two drugs offer a cheaper but far less satisfactory solution to governments that are not prepared to take on the ultra-processed food lobby. CNBC, 31 January 2024, 'Novo Nordisk hits $500 billion in market value as it flags soaring demand for Wegovy, Ozempic'. https://tinyurl.com/NNordisk500bn-Wegovy-Ozempic
56. *Financial Times*, John Gapper, 2021, 'Empire of pain – the story of the Sacklers and OxyContin'. https://tinyurl.com/empirepain-Sacklers-Oxycontin. For the film drama version on Netflix, try *Painkiller*: www.netflix.com/title/81095069

    And to read (without paywall) how this finished up, with the Sackler family retaining $5 billion of their wealth and gaining immunity to further civil (but not criminal) prosecution, try reading 'US court grants Sackler family protection from future

opioid lawsuits in return for $6bn pay out.' *BMJ* 2023, *381*, 1261. https://tinyurl.com/sackler-opioid-payout (published 1 June 2023).
57. Camilo Mora et al. (2022). 'Over half of known human pathogenic diseases can be aggravated by climate change.' *Nature Climate Change*, *12*, 869–875. https://tinyurl.com/clim-change-worsens-disease
58. 'An AI experiment generated 40,000 hypothetical bioweapons in just 6 hours.' https://tinyurl.com/AI-40k-hypothet-bioweapons-6h
59. The first 50 pages of my own book *There Is No Planet B* give a summary of the food and land system. See also the bibliography for more from a variety of angles and in more detail.
60. Misdirecting attention away from the need to reduce fossil fuel use and towards the development of carbon capture and storage (CCS), which is lower priority, is emerging as a key strategy for oil and gas companies intent on maintaining their extraction rate. The London Science Museum ran a Shell-sponsored 'Future Earth' exhibition which focused not on cutting fossil fuel but on CCS. And the Royal Society produced a report on CCS from a working group that it failed to mention had been chaired by the head of the BP Institute, who held the title 'BP Professor'. '*Financial Times*, Royal Society and academics clash over influence of oil and gas industry'. https://tinyurl.com/clash-influence-oilgasindustry . So sad when institutions that should be highly credible try to persuade me that their content is free from the influence of funders with murky agendas. I have seen so many times that this so-called 'Chinese wall' can't be maintained.
61. BBC, Justin Rowlatt, 22 May 2024, 'UK breakthrough could slash emissions from cement'. https://tinyurl.com/cement-slash-emissions

62. David Lee et al. (2021). 'The contribution of global aviation to anthropogenic climate forcing for 2000 to 2018.' *Atmospheric Environment*, 244, 117834. https://tinyurl.com/aviation-radiative-forcing. According to this authoritative paper, non-CO_2 radiative forcing effects are 95% certain to add between 61% and 323% for CO_2 effects, with a central estimate of 194%.

    The largest of the various non-CO_2 radiative forcing effects arises in contrails, which help to reflect sunlight during the day, but at night reflect the Earth's heat back down again. The effects depend on several factors such as altitude (so long-haul is worse), whether flying over land, sea or ice (sea is worse), whether it is day or night (night is worse) and the weather conditions. According to this paper, the best estimate for aviation's non-CO_2 impact is equivalent to emitting 70% more CO_2 when considered over a 100-year period.

63. Our World in Data, Hannah Ritchie, April 2024, 'What share of global CO_2 emissions comes from aviation?' This is a good summing-up of aviation's impacts and trends. https://ourworldindata.org/global-aviation-emissions

64. Intergovernmental Panel on Climate Change. *Global Warming of 1.5 °C* (IPCC, 2018), chapter 4: 'Strengthening and implementing the global response', de Coninck, H., Revi, A., Babiker, M. et al., p. 374. https://tinyurl.com/IPCC-strengthn-glob-response

65. Nuclear risks: it is quite wrong to use death and injury statistics from different energy sources to assess their relative safety. The issue is the near misses. At Chernobyl, we came incredibly close to losing half of Europe. Going forward, trust is a critical issue. As I write this, there is claim and counter-claim at the UK's Sellafield site, near where I live, about the extent of radioactive leakage. Historically, there have been cover-ups – so the default position has to be *not* to trust, unless something has clearly changed in the industry to justify us having renewed confidence in its honesty. I'm not sure what evidence could be provided to signal such a step change, but without knowing we can trust the

integrity of the nuclear industry, we should steer clear. For an interesting account of emerging nuclear waste management techniques, I recommend Yasmin Ali's book, *Powering Up: An Engineer's Adventures into Sustainable Energy* (Hodder, 2024).

66. I wrote about this in *There Is No Planet B*. It comes from a combination of being crowded, northern and cloudy. The four who are worse are Bangladesh, Rwanda, the Netherlands and Belgium. There are some very small further outliers.

67. The International Energy Agency reports, 'Total energy-related CO_2 emissions increased by 1.1% [between 2022 and] 2023. Far from falling rapidly – as is required to meet the global climate goals set out in the Paris Agreement – CO_2 emissions reached a new record high of 37.4 Gt [billion tonnes] in 2023.' However, it goes on to contradict itself by saying 'Clean energy is at the heart of this slowdown in emissions'. To be clear: there is no slowdown! The rate of emissions is rising. They are speeding up. The IEA's executive director, Fatih Birol, goes on to say, 'The clean energy transition is continuing apace and reining in emissions.' This too is nonsense. There is no clean energy transition. The use of both fossil fuel energy and clean energy is increasing. One is not replacing the other, so transition is simply not taking place to any extent.

    IEA, 1 March 2024, 'CO_2 emissions in 2023: A new record high, but is there light at the end of the tunnel?' https://tinyurl.com/IEA-CO2-2023

68. According to Mark Jacobson's book *No Miracles Needed* (Cambridge University Press, 2023), p. 213.

69. Wikipedia, https://en.wikipedia.org/wiki/Submarine_power_cable

70. Next-generation battery developments look set to improve transportation, but not to enable the huge scale-up in storage capacity required to balance the grid; for example, see S&P Global article, 'The future of battery technology'. https://tinyurl.com/future-battery-tech

71. For example, DeSmog, Rachel Sherringham and Clare Carille, 8 December 2023, 'Big meat and dairy delegates triple at COP28'. https://tinyurl.com/3x-big-meat-dairy-COP28
72. Mike Berners-Lee, *How Bad Are Bananas?* (Profile Books, 2020), p. 148.
73. 'Universal basic services' is a concept along the lines of universal basic income, except that instead of giving people the money to have their basic needs met, it may be more effective and reliable to ensure access to those services. UCL Institute for Global Prosperity, 2017, 'Service prosperity for the future: a proposal for Universal Basic Services'. https://tinyurl.com/proposal-univbasicserv
74. The UK COVID-19 enquiry showed that data availability, while helpful, was not always sufficient for high-quality decision-making.
75. For an example of the increasing appetite to get beneath the superficial layer of the crisis, see Merz, J.J., Barnard, P., Rees, W. E. et al. (2023). 'World scientists' warning: the behavioural crisis driving ecological overshoot.' *Science Progress*, *106*. https://tinyurl.com/behaviour-ecolog-overshoot

4 The Middle Layer of the Polycrisis

1. Podcast, *The Rest Is Politics: Leading*: episode 56, 'Bill Gates: Conspiracy theories, AI, and the politicians he most admires'. About 40 minutes into this interview, Alastair Campbell – ignoring the papers emerging in top scientific journals from some of the world's most eminent scientists – introduces the climate with a patronising 'A lot of young people are getting very worried about climate', ignoring the calls from octogenarians such as Sir David King, chief scientific adviser, Lord Martin Rees, former president of the Royal Society, or James Hansen, NASA scientist and so-called 'Godfather of Climate Science'.
2. According to Stack Exchange, this quote – variations of which are often attributed to Einstein – was more likely to have originated in about 1970, 15 years after Einstein's death. It is a good quote,

regardless of who said it first. https://tinyurl.com/change-thinking-quote-who

3. The *Guardian*, 30 November 2023, 'Record revenues at UK gambling firms amid rise of online slot machines'. Dr Matt Gaskell, who runs the NHS Northern Gambling Service, said: '(Our) data confirms that the centrepieces of the gambling industry are the most harmful products. These are rapid, continuous casino products engineered to prolong play, exploit decision-making, and generate unaffordable losses. It's a common issue in our clinics.'

'Generally, the proportion of revenue derived from those with problem gambling ranges from 15–50% across studies.' The National Council on Problem Gambling, November 2019, 'Proportion of revenue from problem gambling'. https://tinyurl.com/GREO-revenue-problem-gamblg

However, profits, compared to revenue, are almost certainly strongly skewed towards problem gambling rather than, for example, national lotteries and other 'charity' gambling. See also endnote 20 in this chapter.

4. Examples are too many to list, but here is one that tells you all you need to know. In March 2024, Conservative Party headquarters released an advert claiming that Muslim London Mayor Sadiq Khan had 'seized power' (in fact he had been fairly elected), and that London had since become a crime centre (when in fact it was one of the safest capital cities in the world and had a crime rate below the UK average). The same video suggested that police enforcing the rules on lower-emissions vehicles were compelling people to stay indoors, and conjured up fearful images of them on the basis that their uniforms were black and they sometimes wear masks (presumably for air filtration). The government had suppressed its own report on Low Traffic Networks, because the report showed that the public actually liked them: see the *Guardian*, Peter Walker, 8 March 2024, 'Rishi Sunak's report finds low-traffic neighbourhoods work and are popular'. https://tinyurl.com/Sunak-rprt-finds-LTEs-popular

The same video also contained footage of a panic in a subway that turned out to be from New York, rather than London. https://tinyurl.com/guardian-Toriesdelete-chaos-ad

See also 'The *Observer* view on the London mayoral video: dog-whistles and lies show Tories will stop at nothing to win', March 2024. https://tinyurl.com/Tory-mayoral-ad-misleading

5. Amnesty International, 14 June 2023. https://tinyurl.com/publicorderlaw-dark-new-era
6. *Byline Times*, Josiah Mortimer, 1 November 2023. https://tinyurl.com/convictedXR-founder-speech
7. The *Guardian*, Sandra Laville, 22 April 2024. https://tinyurl.com/sign-case-thrownout-juryrights
8. The *Guardian*, Jim Waterson, 24 August 2022. https://tinyurl.com/tory-agent-shaping-news
9. The *Guardian*, Ben Quinn and Rowena Mason, 24 February 2024. https://tinyurl.com/GBNews-breaking-impartiality
10. Between 1 April 2023 and 31 March 2024, the Trussell Trust gave out over one million food parcels specifically for children, out of the nearly three million it delivered UK-wide. www.trusselltrust.org/news-and-blog/latest-stats/end-year-stats/
11. The *Independent*, Ella Pickover, 3 July 2023, 'The NHS is "collapsing" – leading medic warns'. https://tinyurl.com/lead-medic-NHScollapsing
12. The UK charity National Energy Action (NEA) estimates recent numbers of households in fuel poverty here: https://tinyurl.com/UK-fuelpoverty-trends
13. Transparency International UK sadly has many examples on its website: www.transparency.org.uk/
14. Trickledown is the idea that you can improve the lot of everyone by making the rich richer and letting their wealth 'trickle down' to the poor. Disproving trickledown theory is simple, and illustrated in my book *There Is No Planet B* (2021 edition), pp. 147–149, 'What is trickledown and why is it dangerous?'

15. This 2022 report highlighted that a new billionaire is minted every 26 hours, while inequality contributes to the death of one person every four seconds: https://tinyurl.com/oxfam-inequality-kills
16. *Heated*, Arielle Samuelson, 24 April 2024, 'Nobel Prize-winning economist calls for climate tax on billionaires'. https://tinyurl.com/Nobel-economist-tax-bnaires
17. Dan Gilbert, *Stumbling on Happiness* (Harper Perennial, 2007). In terms of income, the plateau comes much earlier – $75,000 in 2010, according to Killingsworth, M.A., Kahneman, D. and Sellers, B. (2023). 'Income and emotional well-being: a conflict resolved.' *Proceedings of the National Academy of Sciences.* https://tinyurl.com/income-and-well-being. For an accessible and carefully researched book on this, I recommend *The Spirit Level: Why Equality Is Better for Everyone* by Kate Pickett and Richard Wilkinson (Penguin, 2010).
18. Bloomberg Billionaires Index (as of 14 August 2024).
19. In my view, for example, Jeff Bezos's Earth Fund helps to finance the Science-Based Targets Initiative, but under its influence the SBTi's board of trustees announced that company 'net zero' targets could include a range of offsets, effectively depriving the SBTi target regime of any teeth, and letting big corporations off the hook from having to take meaningful action. Many SBTi staff threw their hands up in horror, but at the time of writing, it is uncertain whether Bezos's money or the employees' integrity will have more sway.

    The *Guardian*, Patrick Greenfield, 20 May 2024, 'The Bezos Earth fund has pumped billions into climate and nature projects. So why are experts uneasy?' https://tinyurl.com/bezos-earth-fund-influence

    Another example is the Sackler family – who have been accused (with robust evidence in my view) of knowingly contributing to the US opioid crisis – funding the arts in the UK and elsewhere. The *Guardian*, Ben Quinn, 9 January 2022,

'Sackler Trust gave more than £14m to UK public bodies in 2020'. https://tinyurl.com/Sackler-14m-to-UKpublicbodies

Denise Coates, who has made billions of pounds from online gambling, a significant proportion of which is from 'problem gambling' (a euphemism for people having their lives ruined through addiction), set up the 'Denise Coates Foundation' (note how these philanthropic funds are so often named after the person founding them) to give away pitiful slivers of her fortune to causes she deems worthy. The *Guardian*, Rob Davies, 2 January 2022, 'Betting billionaire's charity cuts donations from £9m to £6m'. https://tinyurl.com/betting-bnaire-cuts-donations

20. The Oxfam Assembly, 1994, was a three-day gathering of 250 Oxfam stakeholders to debate poverty, its causes and solutions.
21. The *Guardian*, Hamilton Nolan, 21 December 2023. https://tinyurl.com/Zuckerberg-bunker-hawaii
22. The *Guardian*, Douglas Rushkoff, 4 September 2022. https://tinyurl.com/apoc-survival-richest
23. The *Guardian*, Mark O'Connell, 15 February 2018. https://tinyurl.com/SiliconV-bnaires-apoc-NZ
24. There are those who argue that only GDP needs to grow and that this can be done without physical growth. For evidence, some point to a list of countries that have succeeded in 'absolute decoupling' – that is, one can rise while the other falls – between GDP and carbon emissions. However, as we've seen, the dynamics are global not national, and so showing that one part of the system can decouple is not evidence that the system as a whole can do so. At the system level, the evidence is that global GDP and emissions are both rising. Meanwhile, as discussed earlier, the International Energy Agency predicts rising energy demand to 2050. This is physical growth.
25. Wikipedia: Gross Domestic Product, https://en.wikipedia.org/wiki/Gross_domestic_product
26. Charles Eisenstein, *Sacred Economics: Money, Gift & Society in the Age of Transition* (North Atlantic Books, 2021).

27. An example might be found in Exxon's lack of contrition over having hidden its knowledge of climate change for decades. *Scientific American*, 26 October 2015, 'Exxon knew about climate change almost 40 years ago'. https://tinyurl.com/exxon-knew-climchange-1977
28. Kate Raworth, *Doughnut Economics* (Random House, 2018).
29. Podcast, *The Rest Is Politics: Leading*, episode 22, aired 12 June 2023. Available on many podcast platforms.
30. Bank.Green helps people find ethical and sustainable local banks, aiming for a level of customer demand for 'green banking' that will force banks to defund fossil fuels: https://bank.green
31. www.mymothertree.com/. For transparency, my company has a very small stake in MotherTree.
32. 'State repression of environmental protest and civil disobedience: a major threat to human rights and democracy'. Position Paper by Michel Forst, UN Special Rapporteur on Environmental Defenders. https://tinyurl.com/threat-suppress-env-protest

    It is worth quoting a few paragraphs in full:

    The repression that environmental activists who use peaceful civil disobedience are currently facing in Europe is a major threat to democracy and human rights. The environmental emergency that we are collectively facing, and that scientists have been documenting for decades, cannot be addressed if those raising the alarm and demanding action are criminalized for it. The only legitimate response to peaceful environmental activism and civil disobedience at this point is that the authorities, the media, and the public realize how essential it is for us all to listen to what environmental defenders have to say.

    - In the UK, the 2022 Police, Crime, Sentencing and Courts Act enables the police to restrict and even ban 'noisy' or 'disruptive' public assemblies. In addition, the 2023 Public Order Act grants the police extended powers to restrict peaceful protests. It also introduces new criminal offenses that make some forms of protest illegal, such as creating a criminal offense for 'locking-on' (i.e. attaching oneself to another person, to an object or to a building), or even for being 'equipped' for such acts. This means, for instance, that carrying a bike lock in a public space with the intent to attach something, such as a bike, to something else, such as a fence, could be considered illegal. The UK Government's factsheet on the 2023 Public Order Act expressly refers to environmental protests by Extinction Rebellion, Insulate Britain and Just Stop Oil, all peaceful movements, as the reason for passing this law.

...

- In the UK, courts have prohibited environmental protesters from putting forward defences based on 'necessity' or 'proportionality'. They have also forbidden protesters from mentioning climate change, thereby preventing them from explaining the reasons for their protest. Courts have held convicted environmental defenders who disregarded this prohibition in 'contempt of court' and imprisoned them for up to 8 weeks.
- In the UK, a number of environmental defenders have been imprisoned for peaceful protest, including one defender who received a six-month prison sentence for participating for 30 minutes in a slow march; and two others who were sentenced to two years and seven months and three years of prison respectively, for the blocking of a bridge. The two activists were denied the right to challenge their prison sentences before the Supreme Court.
- Also in the UK, in addition to criminal prosecutions by the State, companies, including companies owned or controlled by the UK government, have taken out civil injunctions against environmental protesters without their knowledge. The injunctions list the names of individuals who have been arrested in relation to protests on a public road or motorway in the past, and are also against 'persons unknown' who may take part in a protest on a public road or motorway in the future. The individuals named in the injunction have

been held liable to pay the company's legal costs for obtaining the injunction, even though those individuals had no knowledge that the injunction was being taken out. Moreover, anyone who breaches one of these injunctions is liable for unlimited fines and imprisonment for up to two years. To date, environmental defenders have received prison sentences of between three and six months for being involved in a road protest in breach of a civil injunction. The fines or imprisonment imposed for breach of a civil injunction is in addition to the sentence the protesters may receive for the criminal charges brought against them regarding the same protest.

. . .

The increasingly harsh approach by the courts in a number of countries towards environmental defenders who have engaged in peaceful protest or civil disobedience, including the courts' ready use of measures designed for counter-terrorism and organized crime, sends a highly alarming signal regarding the state of the civic space and the free enjoyment of fundamental human rights.

33. The *Guardian*, 4 April 2023. https://tinyurl.com/clim-activist-sign-charged
34. Good Law Project, 25 September 2023. https://tinyurl.com/100s-copy-activist-sign
35. Good Law Project, 22 April 2024. https://tinyurl.com/HighCt-slaps-down-prosecution
36. OpenDemocracy, Anita Mureithi, 22 February 2023. https://tinyurl.com/climate-mention-irrelevant
37. The *Guardian*, Sandra Laville, 8 March 2023. https://tinyurl.com/lawyer-concern-protest-restric
38. The *Guardian*, Damien Gayle, 15 December 2023, 'Just Stop Oil activist jailed for six months for taking part in slow march'. https://tinyurl.com/JSOactivist-1st-jailed-new-law
39. The *Guardian*, Damien Gayle, 2 January 2024. https://tinyurl.com/protester-happy-in-prison

40. The *Guardian*, Damien Gayle, 18 July 2024. https://tinyurl.com/JSO-sentence

    In contrast, at the time of writing, the longest sentence given for violent rioting, including punching a police officer in the face, was three years. The *Guardian*, Josh Halliday, 7 August 2024. https://tinyurl.com/riot-sentence

41. Michel Forst's End of Mission Statement, in which he writes about how deeply troubled he is by the criminalisation of peaceful protest, can be read here: https://tinyurl.com/un-env-defenders

42. There are many articles detailing the economic cost of climate change. Here are just a few: World Economic Forum, 12 October 2023. https://tinyurl.com/cost-of-climate; the *Guardian*, Oliver Milman, 17 May 2024. https://tinyurl.com/econ-damage-climate; LSE's *Policy Brief*, 'What will climate change cost the UK?' May 2022. https://tinyurl.com/LSE-climate-cost-UK

43. Noam Chomsky, *How the World Works* (Penguin, 2012).

44. European Court of Human Rights, 9 April 2024. www.echr.coe.int/w/grand-chamber-rulings-in-the-climate-change-cases

    Frustratingly, and appallingly, the Swiss government voted 111 to 72 to reject the ruling. The *Guardian*, Ajit Niranjan, 12 June 2024. https://tinyurl.com/Swsreject-ruling-fem-elders

45. Case reference: *R(Friends of the Earth & Ors) v SS Energy and Net Zero* [2024] EWHC 995 (Admin), summarised at https://tinyurl.com/HCt-carb-bdgt-plan-unlawful and full judgement available at https://tinyurl.com/FotE-high-court-judgment

46. The court ruled that young people have a constitutional right to a healthy environment. *New York Times*, David Gelles and Mike Baker, 16 August 2023. https://tinyurl.com/montana-youthclimate-ruling

47. Sky News, Victoria Seabrook, 17 November 2023, 'Extinction Rebellion activists who smashed HSBC windows found not guilty of criminal damage'. https://tinyurl.com/XRsmashwindows-notguilty

48. The *Guardian*, 9 February 2024. https://tinyurl.com/USclimsci-wins-1m-lawsuit

Notes to pages 114–116

49. BBC, 20 June 2024. https://tinyurl.com/oilproj-count-full-clim-impact
50. The Council never formally granted planning permission – the planning committee voted to approve but the permission was never issued because the developer needed to sign a legal agreement; however, the Secretary of State called it in before they had the option. So technically the Council didn't grant permission, even though they resolved to do so (three times)!
51. In 2022, profits were even higher, at a staggering £32 billion. The *Guardian*, Jillian Ambrose, 1 February 2024, 'Shell to raise dividends again despite 30% fall in annual profits'. https://tinyurl.com/Shell-profitsdown-dividendsup
52. Greenpeace, Emily Black, 9 November 2023, 'Shell hits Greenpeace with intimidation lawsuit: threatening $8.6m damages claim and protest ban to silence climate demands'. https://tinyurl.com/Shell-Gpeace-intimid-lawsuit
53. The show-stopping problem with the Science Museum's Shell-sponsored 'Future Earth' exhibition, which made it worse than useless in my view, is that it misdirected attention by focusing primarily on direct air carbon capture and storage (DACCS) with barely a mention of the clear priority, which is, as discussed, to leave the fossil fuel in the ground.
54. West Cumbria Mining issued a highly inaccurate and misleading response to a Green Alliance publication; the response was prepared by a coal mining advocate. Professor Rebecca Willis and I issued a response to the document, which was backed with robust evidence and left almost nothing standing in the advocate's statement. 'Green Alliance report – response to WCM from authors', June 2020. https://tinyurl.com/authors-response-to-WCM
55. Technically, 'fiduciary duty' exists to ensure that those who manage other people's money act in their beneficiaries' best interests. It has become normal to interpret this narrowly as

meaning no more than shareholder profit, regardless of the consequences for employees, the wider public and the planet.
56. Letter from 19 US state Attorney Generals to BlackRock, 4 August 2022. https://tinyurl.com/Letter-19US-AGs-to-BlackRock
57. I witnessed this first-hand when I gave a talk at an asset management conference in Virginia shortly afterwards.
58. For a chilling and in-depth analysis of the current trajectory of AI, I recommend Nate Hagens's interview with Daniel Schmachtenberger on *The Great Simplification* podcast. (He is also spot on in his description of how rebound effects play out in the climate impacts of AI.) 'Silicon dreams and carbon nightmares: the wide boundary impacts of AI', *TGS* 132, recorded 27 June 2024. https://tinyurl.com/wide-bound-ai
59. Sam Bright, Bullingdon Club Britain, March 2023.

5 The Core of the Polycrisis

1. 'Collective Interiority' is the phrase Jonathan Rowson used in 'The Inner Development Goals on trial', published by *The Joyous Struggle*, October 2023. https://tinyurl.com/Rowson-Innerdev
2. Inner Development Goals. www.innerdevelopmentgoals.org/
    An interesting and quirky but serious critique and defence of the IDG framework can be found in Jonathan Rowson's article 'The Inner Development Goals on trial', October 2023. https://tinyurl.com/Rowson-Innerdev

6 Truth – the Single Most Critical Lever

1. *Scientific American*, 26 October 2015, 'Exxon knew about climate change almost 40 years ago'. https://tinyurl.com/exxon-knew-climchange-1977
2. Podcast, *The Rest Is Politics: Leading*, episode 56, aired 22 January 2024. Available on many podcast platforms.

3. Harry Frankfurt, *On Bullshit* (Princeton University Press, 2005).
4. Peter Oborne, *The Assault on Truth*, and www.political-lies.co.uk
5. Peter Oborne, *The Assault on Truth*, pp. 54–57. Each Johnson story I've listed here is in turn referenced. The whole book lists many of Boris's lies, but for an even more comprehensive catalogue, see Oborne's meticulously kept website: www.political-lies.co.uk
6. Senator Bernie Sanders, who is Jewish, insisted in a powerful video on X (formerly Twitter) that it was not antisemitic to hold Benjamin Netanyahu accountable for his actions: https://tinyurl.com/BSanders-X-on-Gaza

    The United Nations Office for Coordination of Humanitarian Affairs gave the following figures for the Gaza–Israel conflict as of 5 April 2024: Palestinian deaths: 33,091 (of whom 35 per cent males over 14 years old, 65 per cent women and children under 14 years old), reported injuries 75,750, predicted to face catastrophic levels of food insecurity 1.1 million, internally displaced 1.7 million. Israelis: 1,200 deaths in Israel, 255 deaths in Gaza, 5,400 injuries, 134 hostages remaining. https://tinyurl.com/impact-GazaIsrael-day181 (accessed 12 April 2024).

    Al Jazeera carried out detailed analysis of the atrocities of 7 October 2023 and found, as we know, horrible crimes, but no evidence to support claims of mass rape and child murder that were made in the *New York Times*, the *Guardian* and many other papers. It reported that in total two babies were killed that day, and that there was no evidence of mass rape: 'The unravelling of the *New York Times* "Hamas rape" story'. https://tinyurl.com/Gaza-toll-6-months-on

    Richard Sanders details the evidence in this Double Down News video: https://tinyurl.com/video-RSanders-Oct7-Israel
7. Jon Ronson, *The Psychopath Test: A Journey Through the Madness Industry* (Picador, 2012). In a similar vein, the 'Dark Triad' of three traits – Narcissism, Machiavellianism and Psychopathy – was coined by Delroy L. Paulhus and found to be fairly common among top management and CEOs. A good overview in Wikipedia: https://en.wikipedia.org/wiki/Dark_triad

7 Getting Truth into Politics

1. The Seven Principles of Public Life (also known as the Nolan Principles) apply to anyone who works as a holder of public office. See the UK government's webpage about them here: https://tinyurl.com/govUK-7principles-public-life
2. See, for example, *Downward Spiral: Collapsing Public Standards and How to Restore Them*, by John Bowers KC (Manchester University Press, 2024). Also, Peter Oborne's *The Assault on Truth* (Simon & Schuster, 2023); Chris Bryant MP's *Code of Conduct* (Bloomsbury Publishing, 2024); Rory Stewart's *Politics on the Edge* (Vintage, 2023); and Caroline Lucas's *Honourable Friends?* (Portobello Books, 2015).
3. Peter Oborne, *The Assault on Truth*, and www.political-lies.co.uk
4. The *Guardian*, 28 April 2015, 'How Thatcher and Murdoch made their secret deal'. https://tinyurl.com/thatcher-murdoch-deal
5. In his book *The Rise of Political Lying*, pp. 91–110, Peter Oborne devotes a whole chapter to Tony Blair: 'The lies, falsehoods, deceits, evasions and artfulness of Tony Blair', in which he contends the CV fabrications are just a small corner of the many untruths that he catalogues.
6. Rory Stewart, *Politics on the Edge*. It is a gripping but depressing account of his journey from standing as MP, to getting elected, rising to cabinet minister, standing to be Prime Minister and finally being ousted for refusing to support a no-deal Brexit. I knew things were bad, but this is still a shock. Why do I trust his account when I'd distrust so many others? That is a very good question to ask. It is because, for me, his track record of integrity, his values, his approach to the work, his record of standing up for his beliefs at the expense of his career, and his life experiences all give his voice a ring of authenticity that is rare among UK politicians.
7. Peter Oborne, *The Assault on Truth*. And www.political-lies.co.uk, which catalogues not just Boris's lies, but those of his successors and other MPs from all parties.

8. Isabel Hardman, *Why We Get the Wrong Politicians* (Atlantic Books, 2018).
9. The *Guardian*, Dan Sabbagh, 15 October 2021, 'Jo Cox, Ian Gow and the UK politicians who have been killed in service'. https://tinyurl.com/UK-politicians-killed
10. *UCL News*, 25 January 2022, 'UK voters value "honesty" most in political leaders'. https://tinyurl.com/votersvaluehonesty
    And the full report on their 2021 survey: *What Kind of Democracy Do People Want?* (The Constitution Unit, 2022).
11. Unlock Democracy and Compass, June 2023, 'UK democracy under strain: democratic backsliding and strengthening 2019–2023'. https://tinyurl.com/UK-democratic-backsliding. This report has a detailed account of declining democratic standards in the UK along with recommendations for reform. It was funded by the Joseph Rowntree Foundation, commissioned by pro-democracy groups, carried out by a credible academic with a strong track record, and concludes with the following 10 recommendations for reform:
    (1) Reform of the Westminster voting system to one in which votes cast are more closely matched to seats won, and so Parliament is more representative of the country as a whole.
    (2) Elections should be made more inclusive by allowing non-photographic voter identification (and 'vouching' at polling stations) and legislating for automatic voter registration to register the millions missing from the electoral roll.
    (3) The independence of the Electoral Commission should be restored.
    (4) Electoral finance legislation should be tightened to prevent 'dark money' being used to influence UK elections.
    (5) Political rights and an appropriate balance of powers should be enshrined in a fully codified constitution.

(6) Power should be more fully and evenly devolved to local communities.

(7) A two-stage reform process should be initiated to deliver reform of the House of Lords – with immediate steps to reduce the house size, introduce greater quality control and remove hereditary peers. The House of Lords should then be reformed to ensure a democratic and effective second chamber.

(8) Citizens' Assemblies should be more fully supported to ensure greater deliberation and involvement in decision-making.

(9) Economic and educational inequalities should be addressed, to take down barriers to democratic participation.

(10) Political literacy and democratic education should be enshrined and resourced in the school curriculum, to enable all schools to create the next generation of engaged citizens.

The *Guardian*, Patrick Butler, 16 March 2023, '"Hostile, authoritarian"' UK downgraded in Civic Freedoms Index'. https://tinyurl.com/UKdowngraded-civic-freedom

12. For example, Rishi Sunak claimed that independent civil service analysts had said Labour would put up taxes by £2,000 per household. The advice turned out to be based on biased Conservative Party assumptions, and Sunak had neglected to mention that the £2,000 figure was not per year but over four years. https://tinyurl.com/claim-Lab-tax-hike2000

Meanwhile, the Conservative Party headquarters once again changed their Twitter/X handle, this time to 'Tax Check UK', to create the false impression that they were independent fact-checkers. https://tinyurl.com/X-taxcheckUK-misleading

13. You can visit Tortoise Media's website landing page via this link: https://tinyurl.com/Tortoise-Westminster-accts

14. According to this 2022 article from *UCL News*, 'UK voters value "honesty" most in political leaders': https://tinyurl.com/UCL-uk-voters-value-honesty

15. In the 1980s, Boris Johnson was sacked by the *Times* newspaper for lying. The *Guardian*, Jamie Grierson, 10 December 2021, 'Lies, damned lies: the full list of accusations against Boris Johnson'. https://tinyurl.com/boris-lies
16. Peter Oborne, *The Assault on Truth*, p. 11.
17. I came at this without knowing much about politics, so I asked around and read around quite a bit. Thanks to Caroline Lucas MP, Rory Stewart, Alastair Campbell and Amber Rudd for conversations. Becky Willis picked over an early draft and suggested some important improvements and tweaks. The following books were particularly influential:
    - *Downward Spiral*, John Bowers KC, 2024
    - *Politics on the Edge*, Rory Stewart, 2023
    - *Code of Conduct: Why We Need to Fix Parliament – and How to Do It*, Chris Bryant MP, 2023
    - *Why We Get the Wrong Politicians*, Isabel Hardman, 2018

    The *Byline Times* and the *Guardian* have been particularly rich sources of details of the problems with the existing system.

    The Governance Project, headed by Dominic Grieve, has produced an interesting and detailed report on Ethics, Codes of Conduct, Parliamentary reform, the Civil Service, Elections and Democracy. www.ukgovernanceproject.co.uk/our-report/
18. The Climate Assembly UK was commissioned by six select committees of the House of Commons to examine the question, 'How should the UK meet its target of net zero greenhouse gas emissions by 2050?' www.climateassembly.uk/

8 Getting Truth into the Media

1. Peter Oborne, *The Assault on Truth*, p. 117. Also the *Guardian*, Nicholas Watt, 21 February 2012, 'Leveson Enquiry has chilling effect on freedom of speech, says Michael Gove'. https://tinyurl.com/leveson-chilling-freedom-spch

2. A 2021 report found that News UK (Murdoch), Daily Mail Group (Viscount Rothermere) and Reach (Richard Desmond) owned a staggering 90 per cent of the UK media. Media Reform Coalition, 30 October 2021. https://tinyurl.com/who-owns-uk-media
3. The *Guardian*, Dana Nuccitelli, 14 July 2014, 'Rupert Murdoch doesn't understand climate change basics, and that's a problem'. https://tinyurl.com/murdoch-doesnt-get-climchge
4. For example, *Byline Times*, Chris Blackhurst, 3 April 2024, 'Telegraph takeover bid backed by UAE doesn't matter – because there's an agenda at every newspaper'. Blackhurst, a former editor of the *Independent*, and journalist for many UK papers, describes an instance of Murdoch telling him to write a piece specifically to smear Mohamed Al-Fayed, as an act of retribution for having cost Murdoch money 'down in Texas'. He goes on to describe how reporters and editors kowtow to the owner's agendas as a routine, without them needing to be in the room.
5. See note above, and also David Yelland's comments to the Leveson Enquiry: https://tinyurl.com/Leveson-transcripts-evid
6. *Byline Times*, Dan Evans and Tom Latchem, 9 April 2024, 'We're all "funding hate": UK government biggest spender on GB News advertising'. https://tinyurl.com/UKgov-top-spender-GBnews-ads
7. The *Guardian*, Jonathan Freedland, 9 August 2024, 'You know who else should be on trial for the UK's far-right riots? Elon Musk'. https://tinyurl.com/musk-ukriots

    Musk's X post can be seen here: https://tinyurl.com/musk-x-civil-war
8. Investopedia, 2024, 'How does Facebook (Meta) make money?' https://www.investopedia.com/ask/answers/120114/how-does-facebook-fb-make-money.asp
9. Facebook was found to be one of the biggest sources of misinformation about COVID-19 during the pandemic (AVAAZ, 2020: https://secure.avaaz.org/campaign/en/facebook_threat_health/) and more recently called out for spreading hate speech and misinformation ahead of India's 2024 election

(The London Story, 2024: https://thelondonstory.org/report/slander_lies_incitement/)

10. *New York Times*, Confessore, 2018, 'Cambridge Analytica and Facebook: the scandal and the fallout so far'. https://www.nytimes.com/2018/04/04/us/politics/cambridge-analytica-scandal-fallout.html

11. For example, Harold Evans, former editor of the *Sunday Times*, made it clear to the Leveson Enquiry into phone hacking that Rupert Murdoch interfered with the content of the paper. OpenDemocracy, Ed Jones, 18 April 2019, 'Five reasons why we don't have a free and independent press and what we can do about it'. https://tinyurl.com/5-reasons-UK-press-not-free. The transcript of the Leveson Enquiry is in the National Archives: https://tinyurl.com/Leveson-transcripts-evid

12. Much of this paragraph is drawn from one *Guardian* article, which in turn draws on many linked sources: the *Guardian*, Dan Sabagh, 2023, 'Power and scandal: how Murdoch drove the UK, US and Australia to the right'. https://tinyurl.com/how-Mdoch-drove-US-UK-Aus-to-R . See also the *Guardian*, Jane Martinson, 24 June 2016, 'Did the Mail and the Sun help swing the UK towards Brexit?' (which includes images of the *Mail* headline, 'If you believe in Britain, vote Leave', and the *Sun* headline, 'Be Leave In Britain').

13. 'Climate change is doing more good than harm' – Matt Ridley's entry on Chartwell Speaker Bureau in 2024. www.chartwellspeakers.com/matt-ridley-climate-change-good-harm/

    The *Guardian*, Bob Ward, 25 January 2016, 'Why are some British newspapers still denying climate change?' https://tinyurl.com/why-some-papers-still-deny

14. The transcript of the Leveson Enquiry is in the National Archives: https://tinyurl.com/Leveson-transcripts-evid

15. *Prospect*, Nick Davies, June 2024, 'The Murdoch Spy Papers'. https://tinyurl.com/murdoch-spy-papers

16. *Prospect*, Nick Davies, June 2024, 'The Murdoch Spy Papers'. https://tinyurl.com/murdoch-spy-papers

News UK were approached for comment on 14 August 2024, but, at the time of writing several weeks later, I have still not received a reply.

17. This incident is well known, but here is one summary of it: *Bylines Scotland*, Martin Roche, 9 April 2024. https://tinyurl.com/undermine-constitution-danger. If you need another example among the many that are available, Bob Ward, director of the London School of Economics' Grantham Institute, details the *Daily Mail*'s truth-defying misrepresentation of climate science and climate scientists here: '*Daily Mail* exploits failing regulatory system to mislead its readers', 17 August 2023. https://tinyurl.com/Mail-exploit-regsystem-mislead

18. Bob Ward, London School of Economics, 27 March 2023, 'The *Daily Mail* is still promoting climate change denial'. https://tinyurl.com/Mail-promoting-climchge-denial. The *Daily Mail* article is here: https://tinyurl.com/Mail-chance-drag-Met-gutter

19. Peter Oborne's *The Assault on Truth*, and www.political-lies.co.uk

20. The *Economist*, 14 March 2024, 'The secret deal that saved the Barclays: Was the twin brothers' business empire built on an ancient fraud?' https://tinyurl.com/ancient-deal-saved-barclays

21. Bob Ward (London School of Economics), 9 January 2019, 'Another humiliation for British climate change deniers and their promoters in the media'. https://tinyurl.com/media-humil-UKclimchge-deniers. Also, Bob Ward, 3 June 2019, 'Charles Moore praises Trump using misinformation about energy'. https://tinyurl.com/Trump-praise-misinfo-energy

    The *Telegraph* still has this piece on its website: 'Donald Trump has the courage and wit to look at "green" hysteria and say: no deal'.

22. The article is here: https://tinyurl.com/trump-couragewit-greenhysteria

23. Founded by notorious climate denier Nigel Lawson, the Global Warming Policy Foundation has been perhaps the UK's most influential so-called think tank propagating misinformation to deny or play down the climate crisis. Its funding sources are

For example: https://tinyurl.com/Lawson-inaccur-climchange-R4

24. BBC, Michael Race, 30 April 2024, '*Telegraph* up for sale after takeover collapses'. https://tinyurl.com/telegraph-takeover-collapse

    The *Guardian*, Steven Morris, 7 June 2023. '"Former estate agents": the strange life of the Barclay twins'. https://tinyurl.com/strange-life-barclay-twins

25. According to the *Guardian*, Richard Desmond attempted to avoid paying tens of millions of pounds in tax on a property deal after intervention from central government. He paid £12,000 to sit next to then cabinet minister Robert Jenrick at a Conservative Party dinner. Jenrick has since admitted 'apparent bias' and under public scrutiny withdrew his approval. The *Guardian*, Jim Waterson, 25 June 2020, 'Richard Desmond: The former porn baron caught in a Tory scandal'. https://tinyurl.com/Desmond-pursue-property-profit

26. *FT* Editorial Code, https://aboutus.ft.com/company/our-standards/editorial-code

27. The *Guardian*, 2020, 'Dominic Cummings thinktank called for "end of BBC in current form"'. https://tinyurl.com/Cummings-called-endcurrentBBC

    Also *Byline Times*, Adam Bienkov and Patrick Howse, March 2024, 'The BBC's road to appeasement'. https://tinyurl.com/bbc-road-to-appeasement

28. The Office of Communications, commonly known as *Ofcom*, is the government-approved regulatory and competition authority for the broadcasting, telecommunications and postal industries of the UK.

29. Jacob Rees-Mogg, throughout 2023 and into 2024, often received more money *per month* from GB News than a full-time minimum-wage worker in the UK would earn in 16 months: www.tortoisemedia.com/westminster-accounts-explore/

30. For example, David Cameron made the licence fee voluntary for over-75s without compensating the BBC and required the BBC to additionally fund the World Service through the same income stream. The licence fee has also not kept pace with inflation. https://tinyurl.com/bbc-road-to-appeasement
31. Tim Davie's first move on appointment was to call the BBC 'too left-wing'. Sam Fowles, *Overruled* (2023), p. 144. And Fowles references a *Telegraph* article about Tim Davie wanting to cut down on left-wing comedy, but I won't direct you to that source that I wouldn't want to spend time viewing, especially as it is behind a paywall that I would not want you to buy into.
32. *Prospect*, Alan Rusbridger, March 2024, 'How the government captured the BBC'.
33. CarbonBrief, 10 August 2017, 'Factcheck: Lord Lawson's inaccurate claims about climate change on BBC Radio 4': https://tinyurl.com/Lawson-inaccur-climchange-R4
34. Dorothy Byrne, Channel 4's head of news, famously called the Prime Minister (Boris Johnson) a 'known liar' and made the point in the 2019 McTaggart Lecture that it was disastrous for a democracy to allow this to happen. https://tinyurl.com/MacTaggartvid-DorothyByrne

    It is interesting to read the various media's responses. The *Telegraph* focuses on her sparking an 'impartiality row', as if the impartial thing to do would have been to let the lies go unchallenged. Here's the link, but it is behind a paywall ... please don't fund the *Telegraph*, because by any serious analysis it fails my criteria. https://tinyurl.com/Telegraph-DByrne-stepsdown
35. The *Independent*, Ivor Gaber, 2017, 'When you actually look into it, the BBC has a centre-right bias': https://tinyurl.com/BBC-centre-R-bias (but it has moved further in the years since then). Ivor Gaber is Professor of Political Journalism at the University of Sussex and was formerly a Westminster-based journalist for the BBC, ITN and Channel Four.

36. This OpenDemocracy article lists think tanks with both E grades and A grades for funding transparency: https://tinyurl.com/think-tank-funding. In my view, those without transparent funding are bogus – they are not being honest with the public about the influential money behind them.
37. The Tax Payers' Alliance website does not list its donors, saying it is a privacy issue – even those donations that come from corporations. At the time of writing (in 2024), it mentions money received in 2020, but not any other year. https://tinyurl.com/TPA-funding. Back in 2009, the *Guardian* ran an article raising concerns over the right-wing nature and connections of the group: the *Guardian*, Robert Booth, 9 October 2009. https://tinyurl.com/who-behind-TPA. DeSmog has written an investigative piece on them, which is also an insightful read: https://tinyurl.com/desmog-TPA
38. The *Guardian* (as part of its Big Money series), Alexander Hertel-Fernandez, Caroline Tervo and Theda Skocpol, 25 September 2018, 'How the Koch brothers built the most powerful right-wing group you've never heard of'. https://tinyurl.com/koch-bros-r-wing-politgroup
39. Civitas has been rated as 'highly opaque' in its funding by Transparify (www.transparify.org) and has been given an E grade for funding transparency by Who Funds You? https://tinyurl.com/civitas-funding
40. They post untruths, such as asserting that climate change is a 'minor global problem'. https://tinyurl.com/civitas-lies
41. See endnote 37 in this chapter.
42. The New Culture Forum does not list its donors on its website. The *Byline Times* has produced an article about its founder, Peter Whittle, which shows his numerous links to right-wing, libertarian movements: The *Byline Times*, Max Colbert, 6 March 2023. https://tinyurl.com/Byline-whittle

43. The Centre for Policy Studies does not list its donors on its website and has received an E grade for transparency from Who Funds You? https://tinyurl.com/CenPolS-funding
44. See endnote 36 in this chapter.
45. There are several articles relating to this, and all are worth reading. DeSmog wrote an exposé on 55 Tufton Street, and the hyperlinks in the article will send you down a murky rabbit hole of shady funding and influence: www.desmog.com/55-tufton-street/

    The *Guardian* has also written about the vested interests supporting the Global Warming Policy Foundation and notes that it has always refused to name its donors: the *Guardian*, Helena Horton and Adam Bychawski, 4 May 2022, 'Climate sceptic think tank received funding from fossil fuel interests'. https://tinyurl.com/climsceptic-ThTnk-fossilfunded

    Mother Jones has written about the influence of neoliberal think tanks in the US, steering policy affecting not only the US population, but the global population, often secretly: https://tinyurl.com/DonorsTrust-dark-money
46. See endnote 36 in this chapter.
47. Daniel Hannon (MEP for SE England) briefly explains this school of thought in a video: https://tinyurl.com/adapt-dont-mitigate
48. The *Guardian*, Felicity Lawrence, David Pegg and Rob Evans, 18 September 2018, 'Right-wing think tanks unveil radical plan for US-UK Brexit trade deal'. https://tinyurl.com/Rwing-plan-US-UK-trade-deal
49. OpenDemocracy, 'Who Funds You?': www.opendemocracy.net/en/who-funds-you/

9 Getting Truth into Business

1. Siddhartha Mukherjee, *The Emperor of All Maladies* (Fourth Estate Ltd, 2011).
2. Naomi Oreskes and Erik Conway, *Merchants of Doubt* (Bloomsbury, 2012). A summary account of the book and its critical acclaim,

including by top scientific journals and other credible sources, can be found on the Wikipedia page: https://tinyurl.com/MerchantsofDoubt-book. As also the *Guardian*, Robin McKie, 1 August 2010, 'A dark ideology is driving those who deny climate change'. https://tinyurl.com/denial-dark-ideology
3. The *Guardian*, Joe Fassler, 3 May 2023, 'Inside Big Beef's climate messaging machine'. https://tinyurl.com/bigbeefPR-climate-messaging
4. Wikipedia, https://tinyurl.com/wiki-UK-PO-scandal
5. For a damning, and moving, exposé of this scandal, including interviews from those affected and shocking legal documents, watch BBC's *Panorama* programme, which aired on 10 January 2024: https://tinyurl.com/bbc-po-scandal

    See also this *Computer Weekly* article from 12 June 2024: https://tinyurl.com/hidden-ITwitness-PO
6. The first of three BBC *Panorama* programmes about the Post Office scandal was aired on 17 August 2015: https://tinyurl.com/bbc-2015-po-scandal
7. Paula Vennells was eventually stripped of her CBE on 23 February 2024.
8. ITV, 2023, 'Mr Bates vs. The Post Office'. Warning: This may bring tears to your eyes and make your blood boil.
9. The *Guardian* has published many recent articles about BAE Systems, available here: https://tinyurl.com/guardian-BAEsystems
10. There is a wealth of compelling evidence against BAE Systems, including their own admission to a US court. US Department of Justice, Office of Public Affairs, 'BAE Systems PLC pleads guilty and ordered to pay $400 million criminal fine'. https://tinyurl.com/BAEguiltyfine

    An investigation into allegations of large-scale corruption in the Al-Yamamah arms deal with Saudi Arabia, including a £20 million slush fund for bribing Saudi princes (with prostitutes, among other things), was contentiously shut down by Tony Blair, on security grounds.

House of Commons Library Research Briefing, 2010, 'Bribery allegations at BAE Systems'. https://tinyurl.com/HofC-bribery-alleg-BAEsys. The House of Lords ruled that the decision to shut down the investigations was lawful, but Lady Hale said, 'I confess I would have liked to be able to uphold the decision ... of the divisional court. It is extremely distasteful that an independent public official should feel himself obliged to give way to threats of any sort.' Campaign Against the Arms Trade, 'BAE Systems: the facts you need to know'. www.caat.org.uk/

Andrew Feinstein's book *The Shadow World; Inside the Global Arms Trade* (Penguin, 2012) details enormous corruption at BAE Systems, including bribery of the ANC to push through a deal. In conversation Feinstein, who was an ANC MP at the time, described BAE Systems to me as one of the three or four worst companies in the world.

See also the *Guardian*: 'The BAE files'. www.theguardian.com/world/bae

For the profits made by BAE Systems from the Ukraine and Israel–Gaza wars, see the *Guardian*, February 2024, 'Arms maker BAE Systems makes record profit amid Ukraine and Israel-Gaza wars'. www.theguardian.com/business/baesystems

11. Here was my polite email, to which I had no response at all, except very indirectly via BAE's head of manufacturing calling the Vice-Chancellor, who asked the head of faculty to meet with me about it.

Dear XXX,

Firstly, apologies for emailing on this special bank holiday. We are hoping to catch you before the meeting tomorrow morning.

Following a great amount of discussion among the team at Small World we regretfully have decided not to undertake this work with BAE Systems. This is a decision that was not taken lightly and has come out of a company-wide clarification on our ethical position, and particularly our commitment to working with an emphasis on truth and honesty, and making the world a better place.

This is a reflection on historical events involving BAE Systems as a whole, and the current importance of exports to the Kingdom of Saudi Arabia for the Air division. We do not have difficulty with the principle of working with an arms manufacturer. However, we are sufficiently uncomfortable about what we think are credible allegations of serious corruption in the 'for profit' sale of arms by BAE, including notably to regimes that are known to be responsible for serious humanitarian violations and possibly war crimes, that we are not comfortable to carry out this work.

I apologise for pulling out at such short notice, and for the inconvenience caused, but we did not want to do so lightly and it has taken us this long to come to a clear position.

We would still be very happy to meet tomorrow, in person or online, to talk through our decision, but not to get involved in the proposed work itself . . .

Best regards
Mike

12. The organisation American Oversight is 'investigating the impacts of anti-ESG bills and the influence networks working behind the scenes', an area of investigation it refers to as a 'threat to democracy': https://tinyurl.com/conserv-attack-ESGinvesting
13. Upton Sinclair, American novelist and social reformer. What he actually said was, 'It is difficult to get a man to understand something when his salary depends on his not understanding it.' I've changed it slightly to include all genders and also to acknowledge that not *everyone* is quite so blinded by their salary.

14. The website for the membership community is accessible here: www.marketingkind.org/
15. BBC News, 5 June 2024, 'Ban fossil fuel ads to save climate, says UN chief'. https://tinyurl.com/ban-fossfuel-ads-says-UN
16. Amnesty International, September 2023, 'Democratic Republic of the Congo: Industrial mining of cobalt and copper for rechargeable batteries is leading to grievous human rights abuses'. https://tinyurl.com/cobalt-copper
17. See my chapter 'How (not) to buy', section 4.20 in Greta Thunberg's *The Climate Book* (Penguin Random House, 2022).
18. The *Guardian*, Ian Gow and Stuart Kells, 3 June 2023, 'The Big Four firms are incapable of unwinding their own deep-seated conflicts'. https://tinyurl.com/big4-cant-unwind-deep-conflict
19. Video by Four Corners, an Australian investigations documentary series: 'The consultancy firms raking in billions of taxpayer dollars'. Broadcast on ABC News, 6 August 2023. https://tinyurl.com/consultFirms-bns-from-taxpayer
20. International Consortium of Investigative Journalists, Michael Hudson, 14 November 2023, 'Beyond Cyprus, PwC has weathered a decade of global probes and scandals'. https://tinyurl.com/beyondCyprus-PWC-scandals
21. The *Guardian*, Julia Kollewe and Kalyeena Makortoff, 12 October 2023, 'KPMG boss says Carillion auditing was "very bad" as firm is fined record £21m'. https://tinyurl.com/KPMG-fined-bad-Carillion-audit
22. Reuters, 3 April 2023, 'EY fined, banned from some audits in Germany over Wirecard scandal'. https://tinyurl.com/EY-fined-audit-ban-Wirecard
23. The *Guardian*, Henry Belot, 14 July 2023, 'Deloitte admits misuse of government information as scandal engulfing PwC widens'. https://tinyurl.com/Deloitte-misused-Aus-govt-info
24. Bayes Theorem says $P(A|B) = P(B|A) \times P(A)/P(B)$ where $P(A)$ and $P(B)$ are the probabilities of events A and B, and $P(A|B)$ and $P(B|A)$ are

the probabilities of A given that B has happened and B given that A has happened. In this case if A represents a big company getting prosecuted for an offence, and B represents them committing an offence, then if we assume that the justice system works perfectly, $P(B/A) = 1$: in other words, if a company gets prosecuted for an offence, then it did commit that offence. In this case $P(B) = 1$ also, because the company did get prosecuted. We might also assume that the perpetrators make an assessment at the time of the offence that $P(A/B)$, the likelihood of getting caught, is low because otherwise they wouldn't take the risk. If so, the actual number of offences committed can be expected to be much higher than the number of times that they get prosecuted. So, the fact that a company has been prosecuted is suggestive (but not proof) of more offences having been committed that we don't know about. If, say, $P(B/A)$ was assessed by the perpetrators to be 10 per cent, then $P(A/B) = 10$, representing a probabilistic expectation of 10 offences committed for every prosecution. All this is just a long-winded way of saying that if big companies only commit big offences based on an assessment that they are likely to get away with it, then it is reasonable to assume there has been more skulduggery under the carpet than has ever surfaced.

25. The *Guardian*, Oliver Milman, 1 February 2023, 'Shell's actual spending on renewables is fraction of what it says, group alleges'. https://tinyurl.com/Shell-mislabels-renew-spends

26. The only legitimate offset for a flight is one that can prove *additionality*. And when you look into it carefully that takes you to only one option: direct air carbon capture and storage (DACCS). Other options such as nature-based solutions are fundamentally finite and need to be done anyway. The price for DACCS is currently a little under \$1,000 per tonne of CO_2, and the companies I've looked into (most notably Climeworks and Carbon Engineering) are having real trouble scaling up and bringing that price down. Others are less well developed still.

There is more on this in the Negative Emissions section of the 2020 updated edition of my book *How Bad Are Bananas? The Carbon Footprint of Everything*.

The Travel Navigator website quotes 590 kg CO_2 for a flight from London to New York, one-way and excluding non-CO_2 radiative forcing. When radiative forcing is taken into account, this takes a return flight to about 3.5 tonnes, costing \$3,500 at \$1,000 a tonne. www.travelnav.com

27. Solid quantification of the damage caused by gambling is very challenging, as this UK government report makes clear: UK Office for Health Improvement and Disparities, 'The economic and social cost of harms associated with gambling in England; Evidence update 2023'. https://tinyurl.com/econsocialcost-gambling-harms

    This report makes an estimate of just the direct impacts on the individual – rather than on those around them who often also have their lives devastated – and makes clear that even for this, its figure of £1.7 billion per year is an underestimate because where things can't be quantified, they have been left out. The figure does include some of the more obvious components such as suicides (as if these can be financially quantified), depression, alcoholism and homelessness. To give an example of the level of omission even where an attempt is made to come to a figure, a very small component of the cost of criminality associated with gambling is included; just the cost of imprisonment but not, for example, the cost of the crimes themselves, or even the judicial and social work processes associated.

    See also endnote 3 of Chapter 4.

28. The *Guardian*, Rob Davies, 2 January 2022, 'Betting billionaire's charity cuts donations from £9m to £6m'. https://tinyurl.com/betting-bnaire-cuts-donations

29. *Business Green*, Stuart Stone, 27 July 2023, 'Analysis: Shell's bumper profits "have not translated into higher renewables investment"'. https://tinyurl.com/Shellprofit-no-boost-renewable

30. *Green Queen*, Tanuvi Joe, 11 June 2021, 'Earth Island sues Coca-Cola over greenwashing claims & false advertisement'. https://tinyurl.com/EarthIsland-sues-CocaCola

   While there are some limitations to Break Free From Plastic's methodology, they are highly transparent about their process, and other organisations have replicated these brand audits using slightly different methods, often arriving at the same or very similar conclusions. It's important to note that in the 2023 audit, PepsiCo technically surpassed Coca-Cola in terms of the total number of branded plastic items identified. However, both in terms of geographic spread (which BFFP prioritises in their ranking) and when adjusting for waste per dollar of revenue, Coca-Cola remains at the top of the list of global plastic polluters. Despite being named the number 1 polluter for six consecutive years, from what I could find, Coca-Cola has not publicly responded to these reports. On the other hand, companies like Nestlé have responded without challenging the evidence or methodology presented by BFFP. Furthermore, I could not find any significant criticisms or challenges to the robustness of BFFP's methodology from other sources. https://tinyurl.com/bffp-2023-report

31. Cleanhub.com, Tamara Davidson, 8 February 2024, 'Greenwashing examples: the nine biggest fines handed out so far'. https://blog.cleanhub.com/greenwashing-examples#five

   Top of the list by a long way is Volkswagen, famously fined $34.69 billion for cynically implementing software to falsify car emissions data. Others in the list are Toyota (holding back emissions data reports), DWS (bogus ESG fund claims), Eni (claiming palm oil diesel was green), Kohl's and Walmart (claiming environmentally friendly bamboo, when it was another material), Goldman Sachs (misleading customers on ESG procedures), Keurig (bogus recyclable coffee pod claims), BNY Melon (overstating ESG credentials of funds), H&M and Decathlon (unsubstantiated claims on product labels).

10 The Evolutionary Challenge and Where Each of Us Fits In

1. James Lovelock, quoted in *Intervention Earth* by Gwynne Dyer (Random House, 2024), p. 213: 'I'd put it somewhere between 100 million and a billion as the number of survivors by the end of the century, if things go on as they're going.' We won't know until the end of the century if James Lovelock is right or not, but the point is that this is serious.
2. Andre, P., Boneva, T., Chopra, F. and Falk, A. (2024). 'Globally representative evidence on the actual and perceived support for climate action.' *Nature Climate Change*. doi:10.1038/s41558-024-01925-3; https://tinyurl.com/world-support-climate-action
3. Same source as endnote 2 above.
4. Jeremy Lent pulls a lot of this together in a highly recommended and widely acclaimed book: *The Web of Meaning: Integrating Science and Traditional Wisdom to Find Our Place in the Universe* (Profile Books, 2021). Lent is a former Silicon Valley CEO who has defected from the tech-centric world view.
5. *Lord of the Flies*, William Golding (Faber and Faber, 1954).
6. *Humankind: A Hopeful History*, Rutger Bregman (Bloomsbury, 2021).
7. I'm not writing a book about whether there is such a thing as free will. I don't think anyone can prove or disprove it. So I'm just taking it as read. If you can't go with that, don't bother reading on.
8. It is important to be wary before borrowing anything from management theory that comes out of an old-school business culture that revolves around a free market economy and isn't Anthropocene-fit. However, the basic idea behind Kotler's strategic planning process is useful. Look at the world out there and what it needs. Then look at what you can offer, and find a way of marrying the two.
9. My book *There Is No Planet B* (2021 revised edition) contains a chapter on protest, and my views on what makes it effective.

Appendix 1 A Taxonomy of Deceit

1. The *Guardian*, Victoria Bekiempis, 17 March 2023, 'Trump's own research showed that voter fraud did not cost him election'. https://tinyurl.com/Trump-debunks-own-voter-claims
2. *Le Monde*, Isabelle Mandraud, 1 October 2023, 'Russia uses lies as weapons of mass destruction'. https://tinyurl.com/Russian-lies-WMD
3. The *Guardian*, Pjotr Sauer, 11 April 2024, 'Wife of jailed British-Russian fears he will meet same fate as Navalny'. https://tinyurl.com/fear-Navalnyfate-jailed-UK-RU
4. This website was originally www.boris-johnson-lies.com but Oborne has renamed it https://political-lies.co.uk since other politicians are worthy of entries: 'Lies, Falsehoods and Misrepresentations from Johnson to Sunak'. Oborne's work is painstaking and robust. His definition of a lie is a little broader than mine, but all or nearly all of his entries deserve a place in my taxonomy of deception. It's a depressing read.
5. The timeline of these comments by Shapps is published on the 'political-lies.co.uk' website: https://tinyurl.com/Shapps-voted-Leave
6. UK Cabinet Office, 25 May 2022, 'Findings of Second Permanent Secretary's Investigation into Alleged Gatherings on Government Premises During Covid Restrictions'. https://tinyurl.com/suegray-report
7. The final straw for Boris Johnson was when it emerged that he had lied to say that when he promoted MP Chris Pincher to Deputy Chief Whip, he had not been aware of Pincher's track record of molesting colleagues. https://en.wikipedia.org/wiki/Chris_Pincher_scandal
8. Channel 4, 3 October 2023, 'Partygate': www.channel4.com/programmes/partygate. This docudrama drawing strongly on the Sue Gray report, and also including real footage from the day, may make you squirm for multiple reasons.

9. Peter Oborne, *The Assault on Truth*, p. 87, and see footnote same page.
10. Peter Oborne, *The Rise of Political Lying* (Simon & Schuster, 2005).
11. CarbonBrief, Daisy Dunne, 8 June 2023, 'Factcheck: Why banning new North Sea oil and gas is not a "Just Stop Oil plan"'. https://tinyurl.com/new-NSea-oilgas-notJSO-plan
12. Flushed out by Full Fact within the week, on 7 February. https://fullfact.org/news/just-stop-oil-funding-labour-pmqs/
13. *Financial Times*, January 2023, 'Why is the NHS in its worst ever crisis?' https://tinyurl.com/2sjfrmdh. Things have become worse in the two years since this piece.

    British Medical Journal, October 2023, 'An NHS under pressure' – 'the health service has been facing years of inadequate planning and chronic underfunding'.
14. The *Spectator*, Dominic Cummings, 9 January 2017, 'Dominic Cummings: how the Brexit referendum was won'. https://tinyurl.com/Cummings-brexref

    If you want to read the article without financially supporting the media source that published so many of Boris Johnson's deceits, you can either subscribe free for a month (don't forget to unsubscribe) or read an article about it in the *London Economic*, 2 August 2017, 'Vote Leave director admits they won because they lied to the public'. https://tinyurl.com/admits-lied-public
15. The Vote Leave campaign was also found to have broken electoral law: https://tinyurl.com/voteLeave-broke-electoral-law
16. Between 2016 and 2020, the NHS budget did increase by £400 million per week in real terms, roughly enough to keep pace with a growing and ageing population. However, the increase did not come from any non-existent Brexit bonus, but rather out of tax increases, borrowing and cuts in other departments. *UK in a Changing Europe*, Mark Dayan, 2023, 'What has Brexit meant for the NHS?' https://tinyurl.com/Brexit-impact-on-NHS
17. A clip of this exchange was published by the *Guardian* in 2022: https://tinyurl.com/DButler-asked2leave-Borislied

18. *Financial Times* article (paywall-restricted access, unfortunately): 'Johnson partygate report approved as Sunak criticised for missing debate'. https://tinyurl.com/Pgate-rprt-approved-Sunak-crit
19. https://tinyurl.com/Sunak-dinner-noVote . Here is an example of something that Oborne calls a lie, but I'll give it the benefit of the doubt and put it in the category of 'not quite proven to be a lie – could be a weak excuse'.
20. The *Independent*, Andrew Feinberg, 22 April 2022, '"I don't recall": Marjorie Taylor Greene evades questions on martial law, "traitors" and alien invasion movies in hearing that could bar her from election'. https://tinyurl.com/MarjTGreene-I-dont-recall
21. The *Independent*, Andy McSmith and Charlie Cooper, 6 July 2016, 'Chilcot Report: Blair didn't tell the truth about WMDs, the deal with Bush or the warnings of fallout – how Britain went to war with Iraq'. https://tinyurl.com/BlairIraq-WMDverdict
22. Claire Perry was Minister of State at the Department for Business, Energy and Industrial Strategy from June 2017 to July 2019.
23. Department for Energy and Net Zero, *UK Energy in Brief 2023*, p. 13, 'In 2020 21.1% of UK energy supply was from low-carbon sources.' https://tinyurl.com/y7wtrcuf
24. Full Fact website. https://fullfact.org/
25. Full Fact, 19 January 2024, 'Overall cost of pandemic support was around £370 billion'. https://tinyurl.com/leadsom-pandemic-support-370bn
26. David Oliver (2023), *BMJ*, *381*, 1259: 'The "40 new hospitals" pledge was always a mirage'. https://tinyurl.com/BMJ-40hospitals-mirage
27. House of Commons Library, Antony Seely, 24 November 2021, 'Research Briefing on Tax Avoidance and Tax Evasion'. https://tinyurl.com/Commons-tax-avoidce-evasion
28. Comedian Joe Lycett did a great exposé on this for Channel 4, and you can read about it here: the *Guardian*, Stuart Jeffries, 24 October 2021, 'Joe Lycett vs. The Oil Giant Review'. https://tinyurl.com/review-joe-lycett-vs-oil-giant

29. Andrew Feinstein, *The Shadow World: Inside the Global Arms Trade* (Penguin, 2012).
30. Rebecca Willis and I, with help from Rosie Watson and Mike Elm, wrote 'The case against new coal mines in the UK', published by the Green Alliance, in 2020: https://tinyurl.com/case-against-new-UK-coal-mines. West Cumbria Mining issued a highly inaccurate and misleading response prepared by a coal mining advocate. We issued a response to this document, backed with robust evidence and which left almost nothing standing in their statement. June 2020, 'Green Alliance report – response to WCM from authors'. https://tinyurl.com/authors-response-to-WCM
31. *Scientific American*, Jim Daley, 16 December 2020, 'Food-industry-backed research gives results funders want, new analysis shows'. https://tinyurl.com/fundedresearch-desiredresults
32. *The Conversation*, article by Stephen Cammiss and Graeme Hayes, 21 February 2023, 'Environmental activists on trial barred from citing climate crisis in their defence'. https://tinyurl.com/trial-barred-cite-climcrisis
33. Peter Oborne, *The Assault on Truth*, p. 93, and see footnote, same page.
34. *Newsnight*, November 2019. James Cleverly refused to acknowledge the mistake when Emily Maitlis repeatedly challenged the 'dystopian' dressing-up of party lines as impartial fact-checking. You can watch the video here: https://tinyurl.com/Tory-Xrebrand-factcheck www.bbc.co.uk/news/technology-50482637
35. Center for American Progress, Jenny Rowland-Shea, 3 March 2022, 'The Biden Administration's easiest climate win is waiting in the Arctic'. https://tinyurl.com/Biden-easy-climate-win-Arctic

    Natural Resources Defense Council, Jeff Turrentine, 13 November 2023, 'Why the Willow Project is a bad idea'. www.nrdc.org/stories/why-willow-project-bad-idea

36. The *Guardian*, 6 January 2021, 'Jenrick criticised over decision not to block new Cumbria coalmine'. https://tinyurl.com/Jenrickcriticised-notblockmine
37. The *Times*, Billy Kenber and Arthi Nachiappan, 6 April 2023, 'Exposed: how Tory MP offered to lobby for gambling investors'. https://tinyurl.com/toryMP-benton-lobby-investig
38. The *Guardian*, Peter Walker, 8 March 2024, 'Rishi Sunak's report finds low-traffic neighbourhoods work and are popular'. https://tinyurl.com/Sunak-rprt-finds-LTEs-popular
39. The *Independent*, Matt Mathers, 11 April 2023, 'Five bits of bad news slipped out by the government before the Easter break'. https://tinyurl.com/Govt-Easter-bury-5xbad-news
40. United Nations, 27 April 2023, 'UN Human Rights Chief urges UK to reverse "deeply troubling" Public Order Bill'. https://tinyurl.com/UN-says-reverse-troubling-bill
41. Home Office, 30 August 2023. https://tinyurl.com/Govt-PublicOrderBill-factsheet

Appendix 2 Honesty and Trust Criteria

1. Full Fact, working backwards from May 2024, 'MPs who have not corrected the record'. https://tinyurl.com/fullfact-MPs-fail2correct
2. Full Fact, 20 February 2023, 'How does the government's minimum service level legislation compare to strike laws in other European countries?' https://fullfact.org/law/minimum-service-strikes-compare-EU/
3. The *Guardian*, Randeep Ramesh, 15 March 2015, 'Grant Shapps admits he had second job as "millionaire web marketer" while MP'. https://tinyurl.com/ShappsMP-2ndjob-millnaire-mktr
4. See Peter Oborne's blistering book *The Rise of Political Lying* (Simon & Schuster, 2005).
5. 'Lies, Falsehoods and Misrepresentations from Boris Johnson to Keir Starmer' (an extensive collection, searchable by person, by party, by other categories . . .). https://political-lies.co.uk/

6. Chris Bryant expresses gratitude to Full Fact for flagging up his error: https://tinyurl.com/ChrisBryant-correctsrecord-onX
7. UK Parliament Register of Members' Financial Interests: https://tinyurl.com/cv6xp9x3. Also see Tortoise Media's Westminster Accounts website: www.tortoisemedia.com/westminster-accounts-explore/
8. For example, Michael Gove attended Murdoch's wedding to Jerry Hall. The *Guardian*, Lisa O'Carroll, 4 March 2016, 'Rupert Murdoch and Jerry Hall wedding to include Brooks and Gove, but not PM'. https://tinyurl.com/Rupert-Jerry-Gove
9. The *Guardian*, Aletha Adu, 14 December 2023, 'MP Scott Benton faces Commons suspension over lobbying scandal'. https://tinyurl.com/Benton-lobbying
10. The *Guardian*, David Leigh, 27 February 2012, 'The schools crusade that links Michael Gove to Rupert Murdoch'. https://tinyurl.com/schools-crusade-gove-murdoch
11. Tortoise Media, Sebastian Hervas-Jones, 11 January 2023, 'Caymans carry-on'. https://tinyurl.com/westminster-accounts-cayman
12. The *Guardian*, Damian Gayle, 2 March 2024, 'Government documents "blow gaping hole" in its case for Cumbrian coalmine'. https://tinyurl.com/Gov-docs-hole-in-coalmine-case
13. *London Economic*, Jack Peak, 11 March 2024, 'Govt scrambles to bury report on LTNs after it concludes they are popular and effective'. https://tinyurl.com/gov-rush-bury-rprt-popularLTNs. This report ran against the government's so-called 'Plan for Drivers' that in Rishi Sunak's words aimed to ' ... prevent Labour politicians from punishing people for simply living their lives' and included no LTNs without local consent.
14. Caroline Lucas, 18 April 2021, 'Letter to the Speaker about PM's lies': https://tinyurl.com/CLucas-Letter-speaker-PMs-lies
15. Clip of the heated exchange involving Dawn Butler in the House of Commons: https://tinyurl.com/DButler-asked2leave-Borislied

Notes to pages 262–269

Appendix 3 What Democratic and Parliamentary Reforms Would Help in the UK?

1. See endnote 17 of Chapter 7.
2. Isabel Hardman's *Why We Get the Wrong Politicians* outlines the relative wealth and opportunity that is generally needed just to enter the race. In the US, things are far worse. Meanwhile, in *Politics on the Edge*, Rory Stewart describes how only freak circumstances made it possible for him to be selected as a Conservative candidate.
3. For example, Rory Stewart's *Politics on the Edge* describes the buffoonery in the Cameron years and the downward trajectory from there. Or try Caroline Lucas's book *Honourable Friends?*
4. *Byline Times*, Josiah Mortimer, 9 April 2024, 'Conservatives nearly double party spending cap just weeks before mayoral elections'. https://tinyurl.com/x2spendcap-before-elections
5. *Politics on the Edge*, Rory Stewart. A gripping and shocking account of Stewart's journey into British politics and his time within. The question I'm left asking is, 'Why did he ever put up with it?'
6. Just to be clear, I've never thought of being in the House of Lords, and never had any inclination to sign up to this service, but I was keen to find out all about it, just to understand its poisonousness.
7. OnePlanet – find out more at www.oneplanet.com
8. OnePlanet's landing page; they say, 'It's time to "connect the dots" on health, climate and equity.' www.oneplanet.com
9. Pooran Desai co-founded Bioregional, created BedZed, a groundbreaking sustainable community development, and set up 'One Planet' communities around the world before moving on to found OnePlanet. I don't usually plug specific tools or companies, but I do think this one is particularly interesting. For transparency, I've come to know Pooran through his sustainability work and consider him a friend. We have discussed major joint projects and even done odd bits together in the past. That's not why I'm writing up OnePlanet here – I'm doing that because it's good!

INDEX

Locators in *italic* refer to figures

acceptance, 96, 97, 221–222
accountability
    media, 158–159
    Nolan Principles, 153, *154*
acid rain, 193
actions checklist, 227–232;
    see also agency
activism/activists, 11
    checklist of actions, 230
    criminalisation of peaceful protest, 101, 111–113
    organisations, 281–283; *see also* Extinction Rebellion; Just Stop Oil
    prevention of jury from hearing motivations of, 246–247
Adam Smith Institute, 190
Advanced Manufacturing Research Centre, UK, 119
advertising industry, 100, 198–200
Advertising Standards Agency UK, 199
agency, personal, 3–4, 13, 212
    checklist of actions, 227–232
    and crisis anxiety, 222–223
    meaningful action, 2
    media use, 191–192
    responsible marketing of goods, 200
    strategic approaches, 225–227, *226*
    systems approaches, 18
    technology, 116–119
    see also activism/activists; challenging dishonesty
agroecology, 74, 83–84; *see also* food production and security
AI *see* artificial intelligence
Aitken, Jonathan, 167
Al Jazeera news network, 274
albedo effect, 50
AMOC *see* Atlantic Meridional Overturning Circulation
An Inconvenient Sequel: Truth to Power film, 294
An Inconvenient Truth film, 294
anger, Kübler-Ross Grief Transition Curve, 96
Anthropocene, 17, 31, *32*, 91
antibiotic resistance, 61, 70, 83, 88–89
anxiety, crisis, 221–223
arms manufacture, 100, 195–196
artificial intelligence (AI)
    business truth and honesty, 205–206
    relationship between humans and technology, 117–118
    thinking skills, 131
    vision for, 123–124
The Assault on Truth (Oborne), 157, 168, 285
Atlantic Meridional Overturning Circulation (AMOC), 50

Index

attention, misdirection of, 208, 244
aviation industry, 66, 77–78, 203–204, 244

B Corp certification, 211
BAE Systems, 195–196
Baldwin, James, 2
Bank.Green, 110, 294
banking, ethical, 110, 294–295
Banking on Climate Chaos, 294
Barclay, Frederick, 184–185
bargaining, Kübler-Ross Grief Transition Curve, 96–97
BBC (British Broadcasting Corporation), 101, 187–189, 270
Before the Flood film, 294
Bellingcat, 279
Benton, Scott, 249, 255
Berners-Lee, Mike
  actions for dealing with Polycrisis, 226
  biases, 12–13
  work on climate change, 4–5
  see also The Burning Question; *How Bad Are Bananas?*; *There Is No Planet B*
biased gathering of evidence, taxonomy of deceit, 245–246
biased selection of evidence, taxonomy of deceit, 246–247
Biden administration, US, 151
big picture, seeing, 1, 6
  actions for dealing with Polycrisis, 225
  decision-making tools, 269
  thinking skills, 130
  see also standing back from the problem

billionaires, 103–105, *104*
  immunity from climate change, 105–106
  media ownership, 176, 184
  see also inequality
biodiversity, 74, 107; *see also* food and biodiversity layer of Polycrisis
biofuels, 66
biological warfare, 72
biomass of wild mammals, humans and livestock, *62*
bird flu, 70, 83
Black, Conrad, 141
BlackRock Inc., 116
Blair, Tony, 140, 157
  criteria for assessing honesty of politicians, 253–254
  CV lies example, 237
  media influence over voters, 181–182
  Strategy Unit, 268
  trustworthiness tests, 259
  weapons of mass destruction issue, 140, 157, 237, 240, 253–254, 259
blogs, reliable resources, 292–293
books, reliable resources, 283–292
boris-johnson-lies.com, 168
Bradbrook, Gail, 226
Break Free From Plastic audit, Coca-Cola, 208
Bregman, Rutger, 220, 289
Brexit campaign, 156, 238–239
  creating false impressions, 248
  lies, 238–239
  media influences, 181–182, 188
  passive dishonesty, 242
Brooks, Rebekah, 183
Bryant, Chris, 254, 286

Bullingdon Club Britain: The Ransacking of a Nation (Bright), 285
bullshit, *141*, *142*
   as blend of fact and fiction, 140–142
   continuum of consciousness, 142, *143*
   deception strategies, 208
   definition, 140
   see also deception/lies and misinformation
bullying
   legal system as tool for, 11, 193
   politicians, 163
The Burning Question: We Can't Burn Half the World's Oil, Coal, and Gas. So How Do We Quit? (Berners-Lee and Clarke), 6
burying bad news, taxonomy of deceit, 250
Bush, George W., 181–182
business
   to benefit humanity, 197–198
   political and media cooperation, 98–102, *99*
   problems of current system, 125
   psychopaths in, 148–149
   to benefit humanity, 197–198
   vision for, 122–123
business truth and honesty, 9–10, 135, 193–198
   AI and tech industry, 205–206
   aviation, 203–204
   challenging dishonesty, 195, 202, 207, 209–210
   company groupthink/culture, 209
   consultancy, 200–201
   ethical businesses, 210–211
   extraction industries for clean energy, 203
   food and farming, 204–205
   fossil fuel industry, 202–203
   gambling industry, 206–207
   greenwashing, 202, 207–208
   marketing and advertising, 198–200
business as usual, 8, 224
But What Can I Do? (Campbell), 287
Butler, Dawn, 239, 258–259
Byline Times, 186–187, 272

calories, *64*
Cameron, David, 157, 183
camouflage, taxonomy of deceit, 250–251
campaigning *see* activism/activists
Campbell, Alastair, 93, 109, 287, 292–293
car industry, 199–200, 208
CarbonBrief, 272
carbon capture and storage, 75, *76*
carbon dioxide emissions, 41–43, *42*, *44*
carbon footprints, 5
   and living standards, 87–88
carbon neutral, greenwashing, 208
carbon offsets, 202, 203, 208, 231
carbon pricing, 26, 73, 79–80
carbon sequestration, 83
Carson, Rachel, 14, 215, 290
Cato Institute, 190
Centre for Policy Studies, 190
The Century of the Self film, 293
CFCs (chlorofluorocarbons), 117
challenges of Polycrisis, 9, 14, *35*
   global systems approaches, 18

habitual responses, 14–15
inner qualities needed, 127–133, *129*, *130*
learning from past mistakes, 15–16
need for temporary pause in human expansion, 17–18
need for wisdom, 16–17
physical challenges, 37
truth and honesty, 18–20
challenging dishonesty, 136, 242
in business, 195, 209–210
fossil fuel industry, 202
gambling industry, 207
media, 174, 192
strength needed, 147–148
change, need for *see* radical change
changes of mind by politicians, valid reasons, 165–166
Channel 4 News, 270–271
charisma, politicians, 153, 216
charisma, psychopaths, 148, 149–150
cherry-picking evidence, 246–247
chicken case study, 61–62
chlorofluorocarbons (CFCs), 117
Churchill, Winston, 247
circular economies, 74, 85–86
cities, comparisons with villages, 30–31
Citizens (Alexander, Conrad and Eno), 289
Citizens' Assemblies/Juries, 121–122, 171–173, 269
Civitas think tank, 190
CJA (Climate Justice Alliance), 282–283
Clarke, Duncan, 6
clean energy transition, 53

Cleverly, James, 248
climate activists *see* activism/activists
Climate Assembly (2020) UK, 172–173
The Climate Book (Thunberg), 290
climate change
author's work on, 4–5
reliable resources, 289–291
symptoms of climate breakdown, 48–49
truth and honesty, 135
see also climate component of Polycrisis
Climate Change Act (2008), 113
Climate Change Committee (CCC), UK, 155
climate change denial
in media, 176, 177, 181, 185, 188
psychopaths, 149
radical changes needed, 213
think tanks, bogus, 189–190
Climate Change: The Facts BBC documentary, 294
climate component of Polycrisis, 41–52, *42*, *43*, *45*
carbon dioxide emissions, 41–43, *42*, 44
methane emissions, *43*, 44
sea ice, *47*
sea temperatures, 47
symptoms of climate breakdown, 48–49
temperature records, 45–48, *46*, *47*
tipping points/positive feedback loops, 49–52
'Climate hysteria' article, 184
Climate Justice Alliance (CJA), 282–283

Climate Majority Project, 282
climate of truth, need for, 136, 234
   mathematical modelling, *145*
climate pledges, national, 26–27
coal mine example *see* West Cumbria Coal Mine
Coates, Denise, 206–207
Coca-Cola, greenwashing, 208
Code of Conduct: Why We Need to Fix Parliament – and How to Do It (Bryant), 286
Colegrave, Stephen, 272
collective interiority, 127
Committee on Standards in Public Life, Nolan Principles, 152–153, *154*
company culture, 209; *see also* business truth and honesty
competence, in politics, 259
competitive market economy *see* free market
congruence, internal, 222
ConocoPhillips, Willow Project, 248
Conservative Party, UK, 147, 151, 153
consultancies
   biased gathering of evidence, 245
   business truth and honesty, 200–201
consumerism
   demands for energy, 53
   ethical businesses, 211
   media influences, 177
   reducing, 74, 85–86, 88
COP (Conference of the Parties) meetings, IPCC, 41–43
Corbyn, Jeremy, 247–248

core of Polycrisis (outdated values and mindsets), 9, 94, 124, 125–127, *126*
   core values, 131–133, *132*
   depth of understanding, 1–2, 16
   Inner Development Goals, 127–128, *129*
   qualities needed, 127–133
   thinking skills, 128–131, *130*
   see also radical change
corporate dishonesty *see* business truth and honesty
Corruption Perceptions Index (CPI), 281
costs, legal, 11, 114–115, 193
Coulson, Andy, 182, 183
COVID-19 pandemic, *42*
   care homes, 236
   Partygate, 195, 236, 239
crisis anxiety, 221–223
critical thinking skills, 131
crop yields, symptoms of climate breakdown, 49
'The Crucial Years' online blog, 292
cruises, luxury
cultural change *see* radical change
Cumbria Coal Mine *see* West Cumbria Coal Mine
Cummings, Dominic, 187, 238–239, 247

DACCS (direct air carbon capture and storage), 75, *76*
Daily Express, 185
Daily Mail, 184
Daily Star, 185
Davie, Tim, 188
Davies, Nick, 182–183

Dawkins, Richard, 220
DDT (dichlorodipheny
    ltrichloroethane), 193
De Abreu, Lucia Whittaker, 112
deception/lies and
    misinformation, 8, 135,
    136–137, 196
  blending fact and fiction, 140,
    141, 142, 143
  Brexit campaign, 238–239
  COVID-19 and UK care homes
    example, 236
  criteria for assessment, 161,
    163, 252–259
  definitions, 138–140
  Donald Trump example, 234
  Grant Shapps voting record
    example, 235–236
  mathematical modelling, 145
  media, 180
  need for consequences, 135,
    140, 196
  Partygate example, 195, 236,
    239
  politics, 234–239
  probable lies, 239–240
  Putin's war in Ukraine
    example, 235
  Rishi Sunak examples, 237, 238
  spotting and exposing, 140,
    143–145, 144
  taxonomy of, 234–240
  think tanks, UK, 135
  Tony Blair's CV example, 237
  types/techniques of, 138–140,
    139, 208
  white lies, 140, 240
  see also bullshit; challenging
    dishonesty; honesty and
    truth; taxonomy of deceit;
    track records
decision-making and problem-
    solving
  best-case scenarios, 224
  tools for UK political reforms,
    268–269
  truth and honesty, 134
Decker, Marcus, 111
Declassified UK, 273
deep fakes
  artificial intelligence, 117
  thinking skills, 131
defence industry, 100, 195–196
definitions
  bullshit, 140
  kindness, 146
  truth and honesty, 137–138
Deloitte consultancy, 201
Democracy Now! radio broadcast,
    273–274
democracy, undermining of,
    111
democratic reforms *see* political
    reforms
Dengue fever, 49, 71
denial, Kübler-Ross Grief
    Transition Curve, 96; *see also*
    climate change denial
Denise Coates Foundation,
    206–207
depression, Kübler-Ross Grief
    Transition Curve, 97
Desai, Pooran, 268–269
DeSmog organisation, 191, 277
Desmond, Matthew, 288
direct action *see* activism/activists
direct air carbon capture and
    storage (DACCS), 75, 76

diseases, 69–72
  dishonesty, passive, 242; *see also* deception/lies and misinformation
'Do the Math' lecture tour, Bill McKibben, 6
documentaries, reliable resources, 293–294
The Donors Trust think tank, 190
Don't Look Up film, 226, 293
Double Down News (DDN), 275
Doughnut Economics (Raworth), 108–109, 287
Dowden, Oliver, 236
Downward Spiral (Bowers), 286
Drilled, 279
Duflo, Esther, 103
dynamics of growth *see* growth dynamics

ecocide, 111, 123
ecology, reliable resources, 289–291
Ecology Building Society, 211
economic growth, 197
  comparison with growth of organisms, 27–31, *28*
  media influences, 177
  middle layer of Polycrisis, 106–109
  problems of current system, 125
  reliable resources, 287–289
  vision for, 123
The *Economist*, 276
economy of scale
  growth of living organisms, 27–30
  limits to growth, *29*
Ecotricity, 211

education
  challenges, 120
  middle layer of Polycrisis, 119–121
  and population growth, 57,
  problems of current system, 125
Eigen, Peter, 281
El Niño effect, 51–52
electoral campaign spending, 264; *see also* political reform
electric cars, greenwashing, 208
empathy, 166
  educational challenges, 120
  in politics, 120
  sociopaths, 108
  thinking skills, 130
The Emperor of All Maladies: A Biography of Cancer (Mukherjee), 193
employment, checklist of actions, 229
energy component of Polycrisis, 52–55, *54*
energy demand
  media influences, 177
  reductions in, 73, 77–80
  scenarios, 53–55, *54*
energy efficiency
  domestic, 53, 77
  human social systems, 30
  and input constraints, 25–26
  Jevons paradox, 22
  rebound effect, 21–25, *23*, 52–55
energy industry, 100
energy prices example, lies, 237
energy transition *see* sustainable energy transition
enjoying life, 223, 232; *see also* wellbeing

Index

environment
  reliable resources, 289–291
  respect for, 131
Environmental and Social Governance (ESG), 110, 197
equilibrium states, living organisms, 29, 29–30
Ernst & Young (EY) consultancy, 201
ESG (Environmental and Social Governance), 110, 197
ethical banking, 110, 294
ethical businesses, 210–211
Ethical Consumer, 295
European Court of Human Rights, 113
evidence-based work, 10
evolutionary challenges *see* radical change
exponential carbon-emissions curve, 5–7, 42
extinction of humans, 213–214, 222–223
Extinction Rebellion, 113, 167, 226, 230, 281–282
extraction industries
  for clean energy, 203
  coal and oil *see* fossil fuel industry

facing up to reality *see* realism
FactCheck, 270–271, 279
fact-checking resources, 276–281
failure to correct errors, 138, 237, 238, 241–242
fake judgement, 249
fake narratives
  social media, 159
  taxonomy of deceit, 249
  see also deep fakes
false balance of opinion, 11
false impressions, deception strategy, 135, 208, 248
farming *see* food production and security
feedback loops *see* tipping points
Feeding Britain (Lang), 291–292
fiduciary duty, investment, 116
films, reliable resources, 293–294
financial information, reliable resources, 294–295
Financial Times, 186, 271–272
first-past-the-post voting system, 158, 263–264
Five Times Faster – Rethinking the Science, Economics and Diplomacy of Climate Change (Sharpe), 290
food and biodiversity layer of Polycrisis, 60–66
  biomass of land mammals, humans and livestock, 62
  calories, 64
  chicken case study, 61–62
  meat-eating, 60, 63–65, 64, 66, 73–74, 82–85
  protein, 65, 66
  sharing the available food, 66
  wastage in the system, 63–65, 66
food industry
  biased gathering of evidence, 246
  business truth and honesty, 204–205
  reliable resources, 289–291
  sustainable food and land transition, 82, 85
food production and security
  reliable resources, 291–292
  symptoms of climate breakdown, 49

fossil fuel industry, 53, 79–80, 100
    law/legal frameworks, 114–115
    marketing and advertising, 199
    truth and honesty, 135,
        193–194, 202–203
Fox News, 176, 179, 181–182
Frankfurt, Harry G., 140
free market, 12–13
    limitations, 108
free will, 18
freedom of society, 111
Fujitsu, Horizon accounting
    system, 194–195
Full Fact, 241, 254, 276–277
funding
    media, 177–178
    MPs, 162
'Future Earth' exhibition, London
    Science Museum, 115
future scenarios
    best-case, 224
    middle layer of Polycrisis,
        121–124
    'small is beautiful', 131
    technological solutions, 89–90
    thinking skills, 130
    see also agency; hope

gambling industry, 100, 206–207
Gary's Economics, YouTube channel,
    293
gas production, North Sea, 155
Gates, Bill, 93, 137
GB News, 101, 187
GDP (gross domestic product)
    carbon footprints, 87–88
    limitations, 107–108
    media influences, 177
    vision for, 123
Gethin, Cressida, 112

Gibb, Robbie, 188
Gingell, Stephen, 111
global cooperation, 89
Global Corruption Barometer
    (GCB), 281
global governance, radical
    changes needed, 223
Global Returns Project, 211
global systems *see* systems
    approaches
global temperature records,
    45–48, 46, 47
Global Warming Policy
    Foundation, 185, 189–190
'Global warming's terrifying new
    math' *Rolling Stone* magazine
    (McKibben), 6
Global Witness NGO, 277
Goldberg, Adrian, 272
Good Energy, ethical businesses,
    211
Good With Money, reliable
    resources, 295
Gore, Al, 294
Gove, Michael, 176, 246, 255, 257
governance, global, 223
Governance Project, 280,
Grantham Institute, London
    School of Economics, 191
'The Great Acceleration', 31, 32
'The Great Simplification' blog, 292
Greenpeace, 115
greenwashing, 96–97, 202,
    207–208
Grieve, Dominic, 280,
gross domestic product *see* GDP
groupthink, 209; *see also* business
    truth and honesty
growth dynamics, human social
    systems, 21, 30–31

Index 359

The Great Acceleration, 31, *32*
modelling of dynamics, 35
need for reset, 32-35, *33*
growth dynamics, living organisms
  comparison with economic growth, 27-31, *28*
  stable/equilibrium states, *29*, 29-30
growth, economic *see* economic growth
The *Guardian*, 185-186, 191, 271
Gulf Stream *see* Atlantic Meridional Overturning Circulation
Guterres, António, 51, 199

habitual responses, challenging, 14-15
Hallam, Roger, 111-112, 226
Hancock, Matt, 236
Harmsworth, Jonathan (Lord Rothermere), 184
Harris, Kamala, 159
health threats, outer layer of Polycrisis, 69-72
heat pumps, 77, 254
heatwaves, symptoms of climate breakdown, 48-49
Hedges, Chris, 274, 275
hidden motives, taxonomy of deceit, 249
High Court of England and Wales, 113
history, rewriting, 247
Hodge, Margaret, 256
honesty and truth, 2, 9, 12-13, 134-136
  artificial intelligence, 117
  best-case scenarios, 224
  in business *see* business truth and honesty
  challenges ahead, 18-20
  core values, 133
  criteria for assessing, *161*, 163, 177-178, 183, 252-259; *see also* track records
  definitions, 137-138
  expectations of honesty, 257-258
  in media *see* media truth and honesty
  need for kindness, 146, 163-164
  Nolan Principles, 153, *154*
  outer layer of Polycrisis, 72
  personal qualities of honesty, 143-145
  in politics *see* political truth and honesty
  psychopaths, 148-150
  rejection of post-truth narratives, 159-160
  respect for, 133
  standing up for, 147-148
  sustainable food and land transition, 83
  see also deception/lies and misinformation
hope, 2, 8-9, 13
  law/legal frameworks, 113-114
  radical changes needed, 216-219
  truth and honesty, 134
hopelessness, 101, 213, 215-216
Horizon accounting system, Fujitsu, 194-195
hospital building example, loopholing, 243

How Bad Are Bananas? The Carbon Footprint of Everything (Berners-Lee), 5, 290–291
HSBC, 113
Huhne, Chris, 183
human social systems, 21, 30–31
    extinction of humans, 213–214, 222–223
    The Great Acceleration, 31, *32*
    modelling of dynamics, 35
    need for reset, 32–35, *33*
    tipping points, 216–219
Humankind: A Hopeful History (Bregman), 220, 289
hydrogen, storage and transportation, 81

IDGs (Inner Development Goals), 127–128, *129*
impact, business, 197
Independent Press Standards Organisation, 185
individual actions *see* agency
industry, circular economies, 85–86
inequality, 160
    middle layer of Polycrisis, 102–105, *104*
    problems of current system, 125
    reliable resources, 287–289
    vision for, 123
    see also billionaires
information sources, assessing honesty of politicians, 162, 256–257
Inner Development Goals (IDGs), 127–128, *129*
input constraints, and energy efficiency, 25–26
Institute for Free Trade, 190
Institute of Economic Affairs, 190
integrity, Nolan Principles, 153, *154*
interconnectedness, 268–269; *see also* joined-up thinking
Intergovernmental Panel on Climate Change (IPCC), 41–43, 155
international cooperation, 89
internet resources, 292–293
investment
    fiduciary duty, 116
    HSBC, 113
    middle layer of Polycrisis, 109–110
    problems of current system, 125
Involve, 283
IPCC *see* Intergovernmental Panel on Climate Change
Iraq War, 140, 157, 237, 240, 253–254, 259
    media influence over, 181–182
Israel, 147

Jarvis, Andy, 5–6
Jay, Paul, 275
Jenrick, Robert, 249
'Jevons paradox', 22
job security, 195, 198, 199, 242
Johnson, Boris, 120, 140, 157, 176
    creating false impressions, 247–248
    criteria for assessing honesty, 254, 258
    lies, 236, 239
    media support of, 181–182, 184–185
    trustworthiness tests, *261*

Index

joined-up thinking, 1–2, 130, 268–269; *see also* thinking skills
joy/enjoying life, 223, 232
Jukes, Peter, 272
Just Stop Oil (JSO), 230, 238, 282

keeping quiet, taxonomy of deceit, 138, 195, 242
kindness, 146
   criteria for assessing honesty of politicians, 163–164
   definition, 146
   political truth and honesty, 163–164, 170
Klynveld Peat Marwick Goerdeler (KMPG), 201
knowledge gaps, 3
Koch, Charles, 189–190
KPMG (Klynveld Peat Marwick Goerdeler), 201
Kübler-Ross Grief Transition Curve, 155
   middle layer of Polycrisis, 94–97, *95*
   radical changes needed, 213, 221–222
   thinking skills, 131
Kuznets, Simon, 107
Kwarteng, Kwasi, 249–250

laboratory foods, 74
Lancaster, Louise, 112
law/legal frameworks
   middle layer of Polycrisis, 110–116
   use in upholding status quo, 11, 193
   vision for, 123
Lawson, Nigel, 188, 189–190

leadership
   best-case scenarios, 224
   Nolan Principles, 153, *154*
Leadsom, Dame Andrea, 241–242
Leapfrog, ethical businesses, 211
learning from past mistakes, 1–2, 15–16
Led By Donkeys campaign group, 278
legal system *see* law/legal frameworks
   lies *see* bullshit; deception/lies and misinformation
lifestyles, sustainable, 230–232
limits to growth, need for reset, 32–35
Limits to Growth (Meadows, Meadows, Behrens and Randers), 14, 215, 287
living organisms, growth of
   comparison with economic growth, 27–31, *28*, *29*
   plot of energy use against body mass, *28*
   stable states/maturity, *29*, 29–30
living standards, and carbon footprints, 87–88
lobbying of ministers, 116, 162, 255, 266
   electoral reforms needed, 265
looking after oneself, 223, 232
loopholes, exploiting, 108, 208, 243
The Lorax (Dr Seuss), 289
Lord of the Flies (Golding), 220
Low Traffic Neighbourhoods (LTNs), 250, 257
Lucas, Caroline, 258–259
Lycett, Joe, 115
Lying (Harris), 285

Mail on Sunday, 184
Major, John, 152
Make My Money Matter campaign platform, 295
Malthus, Thomas, 56–57
Mandela, Nelson, 167
Mann, Michael, 114
marketing
　business truth and honesty, 198–200
　responsibility in, 200
MarketingKind community, 198–200
markets *see* free market
maturity/stable states, living organisms, 29, 29–30
McKibben, Bill, 6, 292
meaningful action, 2
meat-eating, 60, 63–65, 64, 66, 73–74, 82–85
media
　accountability, 158–159
　independence, 157
　cooperation with politics and business, 98–102, 99
　problems of current system, 125
　psychopaths, 148–149
　reliable resources, 270–276
　vision for, 122
Media and Journalism Research Center, 279
media as usual, 8, 224
Media Reform Coalition, 279
media truth and honesty, 9–10, 135, 158–159, 174–177
　avoiding untrustworthy sources, 179
　BBC, 187–189
　Byline Times, 186–187
　checklist of actions, 228
　criteria for assessing trustworthiness, 177–178, 183
　Daily Express/Daily Star, 185
　Daily Mail/Mail on Sunday/Metro/ New Scientist, 184
　Financial Times, 186
　Guardian, 185–186
　podcasts and social media, 179–180
　political and commercial espionage, 182–183
　radical changes needed, 216
　Telegraph/Sunday Telegraph/ Spectator, 184–185
　think tanks, bogus, 189–191
　Times/Sunday Times/Sun/Fox News, 181–182
　what we can do about it, 174, 191–192
Merchants of Doubt: How a Handful of Scientists Obscured the Truth on Issues from Tobacco Smoke to Global Warming (Oreskes and Conway), 193, 285
methane emissions, 43, 44, 45, 50
Metro, 184
microbial resistance, 61, 70, 83, 88–89
middle layer of Polycrisis (deeper issues), 1–2, 9, 16, 91, 92
　billionaires, 103–105, 104
　economic growth, 106–109
　education, 119–121
　immunity from climate change for rich, 105–106
　inequality, 102–105, 104
　investment, 109–110
　Kübler-Ross Grief Transition Curve, 94–97, 95

law/legal frameworks, 110–116
need for politics, media and business cooperation, 98–102, *99*
psychological perspectives, 98
relationship between humans and technology, 116–119
techno-optimism, 92–94, 96–97
vision for, 121–124
Mill, John Stuart, 242
mindsets, outdated *see* core of Polycrisis
mindsets, radical/new *see* radical change
misdirection of attention, 208, 244
misinformation *see* deception/lies and misinformation
Monbiot, George, 275, 291
Moore, Charles, 185
Morris, Philip, 245
Mother Jones organisation, 190, 191
MotherTree, 295
MP Watch, 283
MPs (Members of Parliament)
  conduct, need for reforms, 264–267
  funding, 162
  pay/remuneration, 265–266
  parliamentary code, 264
Munier, Mishuk, 275
Murdoch, Rupert, 157, 175, 176, 179, 181–183, 255
Musk, Elon, 180
mutual respect, in politics, 164–165, 166–167

national climate pledges, 26–27
National Public Radio (NPR), 273

Nationally Determined Contributions (NDCs), 26–27
Nature Climate Change journal, 217
Net Zero, greenwashing, 208
The New Climate War (Mann), 290
New Culture Forum, 190
New Scientist, 184
New York Times, 274
'The News Agents' blog, 292
News International, 182–183
News Literacy Project, 280
News of the World, 182, 183
news, reliable resources, 270–276
Nikkei, 271–272
1984 (Orwell), 247
Nolan Principles of Standards in Public Life, 152–153, *154*, 242
North Sea, oil and gas production, 155
NPR (National Public Radio), 273
nuclear weapons, 117

Obama, Michelle, 132, 267
obesity, 70–71
objectivity, Nolan Principles, 153, *154*
Oborne, Peter, 235, 253–254, 272, 275, 278,
  The Assault on Truth, 157, 168, 285
  The Rise of Political Lying, 157, 253–254, 284
ocean currents *see* Atlantic Meridional Overturning Circulation
Octopus Energy, 211
Odey Asset Management, 249–250
Ofcom, 101
offsets, carbon, 202, 203, 208, 231
oil production, North Sea, 155

Once You Know film, 293
OnePlanet digital, 268–269
online resources, 292–293
OpenDemocracy organisation, 191, 275
openness, Nolan Principles, 153, 154
optimism, radical changes needed, 216–219; *see also* hope
Orwell, George, 247
Oborne, Peter, 235, 253–254, 272, 275, 278,
  The Assault on Truth, 285
  The Rise of Political Lying, 284
outer layer (physical aspects) of Polycrisis, 9, 37–39
  components, 37, 38, 39; *see also* climate component; technological solutions
  diseases and health threats, 69–72
  energy component, 52–55, 54
  food and biodiversity, 60–66
  pollution, 66–69
  population component, 56–60, 57
  Sustainable Development Goals, 39–41, 40
Outrage and Optimism online blog, 293
Overruled (Fowles), 286
ownership criteria to assess media trustworthiness, 177
Ozempic, anti-obesity drug, 70–71
ozone-depleting aerosols, 117, 193
ozone layer, 117, 125

pandemics, 88–89; *see also* COVID-19 pandemic
Paris Agreement, 26–27
parliamentary code for MPs, 264; *see also* political reforms
party politics, transcendence of, 11
Partygate, COVID-19 pandemic, 195, 236, 239
  passive dishonesty, 242; *see also* deception/lies and misinformation
Patagonia, ethical business, 210
peaceful protest, criminalisation, 101, 111–113
pensions, ethical, 295
Perry, Claire, 241
personal actions *see* agency
personal protective equipment (PPE), 236, 250
personal qualities, criteria for assessing honesty of politicians, 163
Perspectiva, 127
perspective-taking *see* big picture; standing back from the problem
pessimism, 215–216; *see also* hopelessness
  pharmaceutical industry, 70–71, 100
philanthropy, billionaires, 104–105
philosophical perspectives
  bullshit, 140
  crisis anxiety, 222–223
  economic growth, 289
  interconnectedness of all things, 213–214
phone hacking, 182–183
photosynthesis, food produced without, 84–85
physical aspects of Polycrisis *see* outer layer

plastic pollution, 67-69, *68*, 208
Plastic Fantastic film, 293
podcasts
    checklist of actions, 228
    media truth and honesty, 179-180
    reliable resources, 292-293
political reforms, 170-171, 262-263
    conduct of MPs, 264-267
    decision-making tools, 268-269
    electoral campaign spending, 264
    public appointments system, 267
    voting systems/electoral reforms, 263-264
political truth and honesty, 9-10, 100-101, 102, 135, 151-152, 153-156, 173
    checklist of actions, 227-228
    Citizens' Assemblies/Juries, 171-173
    criteria for assessing honesty, *161*, 163, 252-259
    dealing with dishonest politicians, 166-167
    deciding who to vote for, 168
    declining standards, 157
    democratic and parliamentary reform, 170-171, 173
    expectations of honesty, 257-258
    kindness, 163-164, 170
    mutual respect, 164-165
    Nolan Principles, 152-153-*154*
    other factors, 163-164
    regaining trust, 167-168
    rejection of post-truth narratives by public, 159-160
    reliable resources, 284-287
    risk of further declining standards, 160
    taxonomy of deceit, 234-239
    track records, 143-145, 162, 252-254
    trustworthiness tests, 259, *260*, *261*
    unsuitability of candidates for MPs, 158-159
    valid reasons for changes of mind in politicians, 165-166
    voting out of self-interest, 169-170
Political-Lies.com (formerly the boris-johnson-lies website), 278,
politics, 11-12
    media influences, 175-176
    need for politics, media and business cooperation, 98-102, *99*
    problems of current system, 125
    psychopaths, 148-149
    transcendence of party politics, 11
    vision for, 121-122
politics as usual, 8, 224
Politics on the Edge (Stewart), 265, 287
Politifact, 280
pollution, 66-69, *68*
Polycrisis, 1, 232-233
    changes of approach for dealing with *see* radical change
    helpful actions, 3-4
    layers, 36
    see also challenges of Polycrisis; climate component of Polycrisis; core of Polycrisis; food and biodiversity layer; middle layer; outer layer

population component of
    Polycrisis, 56–60
  educational opportunities, 57,
  enabling decline, 74, 83, 86–88, 123
  growth-based model, 59–60
  Malthus's ideas, 56–57
  population growth since 1850, 57
  population projections, 58
  sustainable food and land transition, 83
  technological solutions, 86–88
  vision for, 123
positive feedback loops *see* tipping points
Post Office miscarriage of justice, 194–195
Post-Truth: How Bullshit Conquered the World (Ball), 285
Post-Truth (McIntyre), 286
Post Truth: The New War on Truth and How to Fight Back (d'Ancona), 286
post-truth narratives, 159–160
Poverty (Desmond), 288
power-seeking, media influences, 177
Power Up (Ali), 291
PPE (personal protective equipment), 236, 250
The Price of Inequality (Stiglitz), 288
PricewaterhouseCoopers (PwC), 201
private education, 120–121
Private Eye, 276
problem-solving *see* decision-making and problem-solving

profit motive, media influences, 177
Prospect magazine, 183, 275–276
Prosperity Without Growth (Jackson), 289
protein, 65, 66
protest, criminalisation of, 101, 111–113; *see also* activism
psychological perspectives, Polycrisis, 98
The Psychopath Test (Ronson), 148, 149
psychopaths, 148, 166, 220, 224
public appointments system, reforms needed, 267
Public Order Bill (2023) UK, 250–251
Punditfact, 280
Putin, Vladimir, 235
PwC (PricewaterhouseCoopers), 201

qualities, criteria for assessment of politicians, 163
qualities needed to tackle Polycrisis, 127–133

radical change, need for, 1–2, 3, 10, 212–214
  best-case scenarios, 224
  checklist of actions, 227–232
  crisis anxiety, 221–223
  difference between how people think and how they think others are thinking, 218, 219
  finding a role in changing, 225–227; *see also* agency
  global systems approaches, 223–224

is it already too late?, 215
optimistic view, 216–219
pessimistic view, 215–216
selfishness, 220–221
strategic approaches, 225, *226*
willingness to contribute, *218*
Rapporteur on Environmental Defenders, UN, 111
Ravenous (Dimbleby), 291
Raworth, Kate, 108–109, 287
Read, Rupert, 282
The Real News Network (TRNN), 275
realism, 2, 37–38
climate component of Polycrisis, 52
Kübler-Ross Grief Transition Curve, 97
radical changes needed, 221–222
rebound effect, energy efficiency, 22–25, *23*, 52–55
recycling, circular economies, 85–86
Rees, Martin, 214
reforms needed in UK politics *see* political reforms
refugees, 159
Regenesis (Monbiot), 291
regulation, need for, 108, 115–116
Reid, Silas (judge), 111, 113
religious beliefs, 168
renewable energy *see* sustainable energy transition
resets, human growth dynamics, 32–35, *33*
resistance, microbial, 61, 70, 83, 88–89
resources, 270
books, 283–292
campaigning organisations, 281–283
fact-checking and whistle-blowing, 276–281
films/documentaries, 293–294
financial information, 294–295
news media, 270–276
online blogs/channels/podcasts, 292–293
respect
for environment, 131
in politics, 266–267
for truth, 133
respect, universal, 11
core values, 131–132
kindness, 146
refugees, 159
The Rest Is Politics podcast, 93, 109, 292–293
Retraction Watch, 280
rich *see* billionaires
Ridley, Matt, 181
The Rise of Political Lying (Oborne), 157, 253–254, 284
Ronson, Jon, 148, 149
Rosindell, Andrew, 256
Rothermere, Lord (Jonathan Harmsworth), 184
Rowson, Jonathan, 127
Rusbridger, Alan, 183, 188, 275–276

Salvation Army, 210
Santa Fe institute, 28
Scale (West), 283–284
SDGs *see* Sustainable Development Goals
sea ice, 47, 49, 50
sea temperatures, 47
self-interest, voting, 169–170

The Selfish Gene (Dawkins), 220
selfishness, radical changes needed, 220–221
selflessness, Nolan Principles, 152–153, *154*
self-reflection, thinking skills, 131
Shapps, Grant, 155
   criteria for assessing honesty, 253
   energy prices lies, 237
   voting record lies, 235–236
Sharp, Richard, 188
Shaw, Daniel, 112
Shell Global, 115, 244–245
shipping, 78
Silent Spring (Carson), 14, 215, 290
Sinclair, Upton, 198
'small is beautiful' vision for future, 131
Small World Consulting, 8, 98, 197–198
social media
   checklist of actions, 228
   fake narratives, 159
   media truth and honesty, 179–180
   relationship between humans and technology, 118
social systems *see* human social systems
social tipping points, 216–219
societal change *see* radical change
sociopaths, 108, 224
solar panels, 84–85
space tourism, 17–18
speaking out *see* challenging dishonesty
The *Spectator*, 184–185
spin, 135
spirit of decency, 108
The Spirit Level (Wilkinson and Pickett), 288
spiritual poverty, billionaires, 105
stable/equilibrium states, living organisms, 29, 29–30
standing back from the problem, 1–2, 9, 16, 21
   economic growth compared with living organisms, 27–31, *28*
   energy efficiency and input constraints, 25–26
   energy efficiency rebound effect, 21–25
   national climate pledges, 26–27
   see also big picture; learning from past mistakes
Stanford Prison Experiment, 220
staying quiet, taxonomy of deceit, 138, 195, 242
steel industry, 77
Stewart, Rory, 109, 157, 265, 287, 292–293
Stiglitz, Joseph, 288
Stockholm Resilience Centre, 31
storage of renewable energy, 73, 80–81
strategic approaches to Polycrisis, 225, *226*
Strategy Unit, UK, 268
strength, standing up for truth, 147–148
subtle twists, taxonomy of deceit, 241
The *Sun*, 176, 181–182
Sunak, Rishi, 155, 157, 160, 188
   criteria for assessing honesty, 253

lies, 237, 238
Low Traffic Neighbourhoods review, 250, 257
probable lies, 239
Sunday Telegraph, 184-185
Sunday Times, 181-182
super-rich *see* billionaires
supermarket chains, misdirection of attention, 244
Sustainable Development Goals (SDGs), UK, 39-41
  core of Polycrisis, 127, *129*
  outer layer of Polycrisis, *40*
sustainable energy transition, 53, 75-81, *76*
  storage, 73, 80-81
  supply, 73, 75-81, *76*
  technological solutions, 75-81, *76*
  transportation, 73, 80-81
sustainable food and land transition, *82*, 85
sustainable lifestyles, checklist of actions, 230-232
Sweeney, John, 272
Switzerland, climate inaction, 113
systems approaches, 6-7
  challenges ahead, 18
  radical changes needed, 223-224
  reliable resources, 283-284
  thinking skills, 130

The Tax Payers' Alliance think tank, 189, 190
taxation system, 12-13, 243
taxonomy of deceit, 140, 153, 234, 251
  altering history, 247
  biased gathering of evidence, 245-246
  biased selection of evidence, 246-247
  burying bad news, 250
  camouflage, 250-251
  creating false impressions, 248
  failure to correct errors, 241-242
  fake narratives, 249
  hidden motives, 249
  keeping quiet, 242
  lies, 234-239
  loopholing, 243
  misdirection of attention, 244
  probable lies, 239-240
  subtle twists, 241
Taylor-Greene, Marjorie, 240
tech industry, business truth and honesty, 205-206
technical fixes, 221-222
technological solutions, 3, 7
  circular economies, 85-86
  evolution of wisdom, 16-17
  middle layer of Polycrisis, 116-119
  outer layer of Polycrisis, 37, 72, 89-90
  population decline, enabling, 86-88
  problems of current system, 125
  relationship between humans and technology, 116-119
  sustainable energy transition, 75-81, *76*
  sustainable food and land transition, *82*, 85
  top-10 issues, 73-75
  vision for, 89-90, 123-124

techno-optimism, 92–94, 96–97, 125, 289
The *Telegraph*, 184–185
telephone hacking, 182–183
temperature records, global, 45–48, *46*, *47*
Thatcher, Margaret, 157, 181–182
There Is No Planet B: A Handbook for the Make or Break Years (Berners-Lee), 7–8, 17–18, 87, 128, 284
'TheyWorkForYou' website, 277–278
things you can do *see* agency
think tanks, UK
   bogus, 189–191
   misinformation, 135
thinking skills, 128–131, *130*; *see also* joined-up thinking; radical change
third-party attitudes, media trustworthiness, 178
Thunberg, Greta, 225, 230, 290
The *Times*, 176, 181–182, 255
Timpson, ethical businesses, 210
tipping points/positive feedback loops
   climatic, 49–52
   social, 216–219
Titanic analogy, deception, 138, *139*
tobacco industry, 193
Tortoise Media, 162, 254, 278
track records
   assessing honesty of politicians, 143–145, 162, 252–254
   James Cleverly, 248
   media sources, 175, 176, 178
   Tony Blair, 237
   unsuitability for responsibility, 156, 167

Trajectory of the Anthropocene, 31, *32*
transparency
   criteria for assessing media trustworthiness, 178
   criteria for assessing politicians, 162, 174, 183, 192, 254–255
Transparency International (TI), 281
transport systems, 77–79, 208
transportation of renewable energy, 73, 80–81
trickledown economics, 12, 102
Triodos Bank, 211
Trowland, Morgan, 111
TRNN (The Real News Network), 275
Trump, Donald, 1, 8, 138, 151
   criteria for assessing honesty, 258
   lies, 234
   media support of, 180–181, 185
Truss, Liz, 157
trust, regaining in politicians, 167–168
trustworthiness, media sources, 176
trustworthiness tests, political truth and honesty, 259, *260*, *261*; *see also* track records
truth *see* honesty and truth

Ukraine war, 160, 235
ultra-processed food, 70–71, 100
undermining of democracy, 111
United Kingdom (UK)
   Advanced Manufacturing Research Centre, 119
   peaceful protest, criminalisation, 111–112
   race riots, 180

universal respect *see* respect
Unlock Democracy, 283
United States (US), 70, 71, 158, 175–176, 180, 181, 197, 216–217, 248, 264
U-turns, valid political reasons, 165–166

values, core, 131–133, *132*, 212
  consultancy, 201
  criteria for assessing media trustworthiness, 178
  media influences, 174–175
  see also core of Polycrisis; radical change
vested interests, in politics, 101, 102, 162, 255–256
villages, comparisons with cities, 30–31
visions for future *see* future scenarios
volunteering, checklist of actions, 230
Vote Leave campaign *see* Brexit campaign
voting out of self-interest, 169–170
voting systems, first-past-the-post, 158, 263–264

Warner, Trudi, 111
wastage, food, 63–65, 66, 83
wealth, and carbon footprints, 87–88
weapons of mass destruction issue, 140, 157, 237, 240, 253–254, 259
The Web of Meaning (Lent), 284
Wegovy, anti-obesity drugs, 70–71
wellbeing, 107
  and GDP, 107–108
  growth of, 107
  and wealth relationship, 103
West, Geoffrey, 28, 30–31, *33*, 33–34, 283–284
West Cumbria Coal Mine, 114–115, 257
  biased gathering of evidence, 245
  biased selection of evidence, 246
  fake narratives, 249
  greenwashing, 208
  misdirection of attention, 244
  what we can do about it *see* agency
whipping system, electoral reforms needed, 264–265
whistle-blowing, resources, 276–281
white lies, 140, 240
Why We Can't Afford the Rich (Sayer), 288
Why We Get the Wrong Politicians (Hardman), 158, 286
Wild Fell (Schofield), 292
wildfires, symptoms of climate breakdown, 48
willingness to contribute (WTC), *218*
Willis, Rebecca, 245
Willow oil drilling project, 248
wisdom, need for, 16–17
work, checklist of actions, 229

X (formerly Twitter), 180, 247–248
xenophobia, media influences, 177

Yelland, David, 181
yield of crops, symptoms of climate breakdown, 49

work with BAE Systems. This is a decision that was not taken lightly and has come out of a company-wide clarification on our ethical position, and particularly our commitment to working with an emphasis on truth and honesty, and making the world a better place.

This is a reflection on historical events involving BAE Systems as a whole, and the current importance of exports to the Kingdom of Saudi Arabia for the Air division. We do not have difficulty with the principle of working with an arms manufacturer. However, we are sufficiently uncomfortable about what we think are credible allegations of serious corruption in the 'for profit' sale of arms by BAE, including notably to regimes that are known to be responsible for serious humanitarian violations and possibly war crimes, that we are not comfortable to carry out this work.

I apologise for pulling out at such short notice, and for the inconvenience caused, but we did not want to do so lightly and it has taken us this long to come to a clear position.

We would still be very happy to meet tomorrow, in person or online, to talk through our decision, but not to get involved in the proposed work itself . . .

> Best regards
> Mike

12. The organisation American Oversight is 'investigating the impacts of anti-ESG bills and the influence networks working behind the scenes', an area of investigation it refers to as a 'threat to democracy': https://tinyurl.com/conserv-attack-ESGinvesting
13. Upton Sinclair, American novelist and social reformer. What he actually said was, 'It is difficult to get a man to understand something when his salary depends on his not understanding it.' I've changed it slightly to include all genders and also to acknowledge that not *everyone* is quite so blinded by their salary.

14. The website for the membership community is accessible here: www.marketingkind.org/
15. BBC News, 5 June 2024, 'Ban fossil fuel ads to save climate, says UN chief'. https://tinyurl.com/ban-fossfuel-ads-says-UN
16. Amnesty International, September 2023, 'Democratic Republic of the Congo: Industrial mining of cobalt and copper for rechargeable batteries is leading to grievous human rights abuses'. https://tinyurl.com/cobalt-copper
17. See my chapter 'How (not) to buy', section 4.20 in Greta Thunberg's *The Climate Book* (Penguin Random House, 2022).
18. The *Guardian*, Ian Gow and Stuart Kells, 3 June 2023, 'The Big Four firms are incapable of unwinding their own deep-seated conflicts'. https://tinyurl.com/big4-cant-unwind-deep-conflict
19. Video by Four Corners, an Australian investigations documentary series: 'The consultancy firms raking in billions of taxpayer dollars'. Broadcast on ABC News, 6 August 2023. https://tinyurl.com/consultFirms-bns-from-taxpayer
20. International Consortium of Investigative Journalists, Michael Hudson, 14 November 2023, 'Beyond Cyprus, PwC has weathered a decade of global probes and scandals'. https://tinyurl.com/beyondCyprus-PWC-scandals
21. The *Guardian*, Julia Kollewe and Kalyeena Makortoff, 12 October 2023, 'KPMG boss says Carillion auditing was "very bad" as firm is fined record £21m'. https://tinyurl.com/KPMG-fined-bad-Carillion-audit
22. Reuters, 3 April 2023, 'EY fined, banned from some audits in Germany over Wirecard scandal'. https://tinyurl.com/EY-fined-audit-ban-Wirecard
23. The *Guardian*, Henry Belot, 14 July 2023, 'Deloitte admits misuse of government information as scandal engulfing PwC widens'. https://tinyurl.com/Deloitte-misused-Aus-govt-info
24. Bayes Theorem says $P(A/B) = P(B/A) \times P(A)/P(B)$ where $P(A)$ and $P(B)$ are the probabilities of events A and B, and $P(A/B)$ and $P(B/A)$ are

the probabilities of A given that B has happened and B given that A has happened. In this case if A represents a big company getting prosecuted for an offence, and B represents them committing an offence, then if we assume that the justice system works perfectly, $P(B/A) = 1$: in other words, if a company gets prosecuted for an offence, then it did commit that offence. In this case $P(B) = 1$ also, because the company did get prosecuted. We might also assume that the perpetrators make an assessment at the time of the offence that $P(A/B)$, the likelihood of getting caught, is low because otherwise they wouldn't take the risk. If so, the actual number of offences committed can be expected to be much higher than the number of times that they get prosecuted. So, the fact that a company has been prosecuted is suggestive (but not proof) of more offences having been committed that we don't know about. If, say, $P(B/A)$ was assessed by the perpetrators to be 10 per cent, then $P(A/B) = 10$, representing a probabilistic expectation of 10 offences committed for every prosecution. All this is just a long-winded way of saying that if big companies only commit big offences based on an assessment that they are likely to get away with it, then it is reasonable to assume there has been more skulduggery under the carpet than has ever surfaced.
25. The *Guardian*, Oliver Milman, 1 February 2023, 'Shell's actual spending on renewables is fraction of what it says, group alleges'. https://tinyurl.com/Shell-mislabels-renew-spends
26. The only legitimate offset for a flight is one that can prove *additionality*. And when you look into it carefully that takes you to only one option: direct air carbon capture and storage (DACCS). Other options such as nature-based solutions are fundamentally finite and need to be done anyway. The price for DACCS is currently a little under $1,000 per tonne of CO_2, and the companies I've looked into (most notably Climeworks and Carbon Engineering) are having real trouble scaling up and bringing that price down. Others are less well developed still.

There is more on this in the Negative Emissions section of the 2020 updated edition of my book *How Bad Are Bananas? The Carbon Footprint of Everything*.

The Travel Navigator website quotes 590 kg CO_2 for a flight from London to New York, one-way and excluding non-CO_2 radiative forcing. When radiative forcing is taken into account, this takes a return flight to about 3.5 tonnes, costing \$3,500 at \$1,000 a tonne. www.travelnav.com

27. Solid quantification of the damage caused by gambling is very challenging, as this UK government report makes clear: UK Office for Health Improvement and Disparities, 'The economic and social cost of harms associated with gambling in England; Evidence update 2023'. https://tinyurl.com/econsocialcost-gambling-harms

    This report makes an estimate of just the direct impacts on the individual – rather than on those around them who often also have their lives devastated – and makes clear that even for this, its figure of £1.7 billion per year is an underestimate because where things can't be quantified, they have been left out. The figure does include some of the more obvious components such as suicides (as if these can be financially quantified), depression, alcoholism and homelessness. To give an example of the level of omission even where an attempt is made to come to a figure, a very small component of the cost of criminality associated with gambling is included; just the cost of imprisonment but not, for example, the cost of the crimes themselves, or even the judicial and social work processes associated.

    See also endnote 3 of Chapter 4.

28. The *Guardian*, Rob Davies, 2 January 2022, 'Betting billionaire's charity cuts donations from £9m to £6m'. https://tinyurl.com/betting-bnaire-cuts-donations

29. *Business Green*, Stuart Stone, 27 July 2023, 'Analysis: Shell's bumper profits "have not translated into higher renewables investment"'. https://tinyurl.com/Shellprofit-no-boost-renewable